Thank you for returning
your books on time.

Abington Free Library
1030 Old York Road
Abington, PA 19001

ALSO BY BRIAN HAYES

*Infrastructure: A Field Guide
to the Industrial Landscape*

Group Theory in the Bedroom,

and

Other Mathematical Diversions

Group Theory in the Bedroom,

and

Other Mathematical Diversions

BRIAN HAYES

Hill and Wang
A division of Farrar, Straus and Giroux
New York

Hill and Wang
A division of Farrar, Straus and Giroux
18 West 18th Street, New York 10011

Distributed in Canada by Douglas & McIntyre Ltd.
Printed in the United States of America
First edition, 2008

The essay "Clock of Ages" was originally published
in *The Sciences*, the magazine of the New York
Academy of Sciences. All other essays originally
appeared in *American Scientist*, the magazine
of Sigma Xi, The Scientific Research Society.

Library of Congress Cataloging-in-Publication Data
Hayes, Brian.
Group theory in the bedroom, and other mathematical
diversions / Brian Hayes.—1st ed.
p. cm.
ISBN-13: 978-0-8090-5219-6 (hardcover : alk. paper)
ISBN-10: 0-8090-5219-9 (hardcover : alk. paper)
1. Technology   2. Science.   I. Title.

T185.H39 2008
500—dc22

2007023337

Designed by Brian Hayes

www.fsgbooks.com

1   3   5   7   9   10   8   6   4   2

For all my editors

Contents

Preface

You walk onto the stage with a violin in your hand, and the audience goes quiet as the conductor nods in your direction. Then you realize you have not rehearsed the Paganini. As a matter of fact, you've never played the violin in your life.

Variations on this classic dream have haunted my nights for years, and once I lived through a waking version of it. In the early 1980s *Scientific American* magazine, where I was working as an editor, wanted to launch a new monthly column called Computer Recreations, a department "concerned with the pleasures of computation." I volunteered to write it. The first thing I had to do was go out and buy a computer, because I had never laid hands on one.

The weeks that followed brought moments of sweaty-palmed tension and doubt, but the eventual outcome was not the nightmare I had feared. One reason is that I got a lot of help, including patient tutoring from some legendary wizards of the computing community, as well as advice from Martin Gardner and Douglas Hofstadter, whose columns had occupied the same place in the magazine in years gone by. Moreover, I discovered that the computer is not like the violin: it doesn't take inborn genius or a lifetime of practice to get sweet music out of it. On the contrary, the computer is a peculiarly forgiving instrument that not only amplifies our talents but also compensates for our flaws and failings. At a trivial level, we see this every day when a word processor corrects our spelling, or when Google fills a lapse in memory. In deeper ways, too, the computer serves as an aid to understanding, exploration, and problem solving.

My gig as the Computer Recreations columnist at *Scientific American* lasted only a few months; I had to return to my editorial duties. But that brief flirtation with writing about computing and mathematics made a lasting impression; when I left the staff of the magazine a year later, I was determined to rekindle the romance. I wrote a number of pieces for a magazine called *Computer Language* and then several more for *The Sciences*, the magazine of the New York Academy of Sciences. Sadly, both of those publications are now deceased. Since 1993 I have been writing a column called Computing Science for *American Scientist*, the magazine of Sigma Xi, The Scientific Research Society. "Clock of Ages," the first chapter in this collection, appeared in *The Sciences*; all the rest of the essays were first published in *American Scientist*.

The topics included here range from deadly serious (war and peace, wealth and poverty) to utterly frivolous (the mathematics of mattress flipping). Two of the pieces look back to earlier eras when computers were built out of brass gears instead of silicon chips. One essay describes imaginary genetic codes that seem far more elegant (to my taste, anyway) than nature's own scheme for interpreting DNA. Yet another chapter takes up the surprisingly tricky question of what it means for two things to be equal. The slogan under which I began—"the pleasures of computation"—still seems an apt description of what it's all about.

The essays cover a time span of a decade—and some of them show their age. At the end of each chapter I have appended a section of "afterthoughts" to bring the discussion up to date. Within the text of the essays I have tried to restrain myself from making extensive changes, but the truth is, I am a notorious itchy pencil; I can't keep my hands off a manuscript. Various minor errors have been silently corrected, and in several places I couldn't resist the temptation to improve a paragraph or clarify a passage. Major goofs—and there are a few of them—are explained in the afterthoughts.

Each of these essays should be seen as a collaboration between the author and an editor. I particularly want to acknowledge the contributions of Peter Brown at *The Sciences* (he is now at *Scientific American*) and Rosalind Reid, David Schneider, and Fenella Saunders at *American Scientist*. (I also thank those magazines for their cooperation in allowing the reprinting of this material.) Joseph Wisnovsky, a friend and colleague of thirty years, presided over the publication of this book. Finally, all of my work owes a debt to the late Dennis Flanagan, editor of *Scientific American* through all its best years and mentor to a generation of science writers.

Group Theory in the Bedroom,

and

Other Mathematical Diversions

CHAPTER 1

Clock of Ages

December 1999. As the world spirals on toward 01-01-00, survivalists are hoarding cash, canned goods, and shotgun shells. It's not the Rapture or the Revolution they await, but a technological apocalypse. Y2K! The lights are going out, they warn. Banks will fail. Airplanes may crash. Your VCR will go on the blink. Who could have foreseen such turmoil? Decades back, one might have predicted anxiety and unrest at the end of the millennium, but no one could have guessed that the cause would be an obscure shortcut written into computer software by unknown programmers of the 1960s and '70s. To save a few bytes of computer memory, they left room for only the final two digits of the year.

We now know that civilization did not collapse on January 1, 2000. Y2K was a nonevent. Nevertheless, in hindsight those programmers do seem to have been pretty short on foresight. How could they have failed to look beyond year 99? But I give them the benefit of the doubt. All the evidence suggests they were neither stupid nor malicious. What led to the Y2K bug was not arrogant indifference to the future. ("I'll be retired by then. Let the next shift fix it.") On the contrary, it was an excess of modesty. ("There's no way *my* code will still be running thirty years from now.") The programmers could not envision that their hurried hacks and kludges would become the next generation's "legacy systems."

Against this background of throwaway products that somebody forgot to throw away, it may be instructive to reflect on a computational device built in a much different spirit. This is a machine carefully crafted for Y2K compliance, even though it was manufactured at a time when the millennium was still a couple of lifetimes away. As a matter of fact, the computer is equipped to run through the year 9999, and perhaps even beyond with a simple Y10K patch. This achievement might serve as an object lesson to the software engineers of the present era. But I am not quite sure just what the lesson is.

A Glory of Gears

The machine I speak of is the astronomical clock of Strasbourg Cathedral, built and rebuilt several times over the past six hundred years. The present version is a nineteenth-century construction, still ticking along smartly at age 160-something.

The Strasbourg Cathedral clock is not a tower clock, like Big Ben in London, meant to broadcast the hours to the city. Although it does have a face on the exterior of the building—a rather undistinguished one that would look more at home on a train station—the main body of the clock is inside the cathedral. And yet it is certainly not a clock you would put on the mantel or hang on the wall. It has a case of carved stone and wood that stands fifty feet high and twenty-four feet wide, with three ornamented spires and a gigantic instrument panel of dials and globes, plus paintings, statuary, and a large cast of performing automata. Inside the clock is a glory of gears.

"Clock" is hardly an adequate description. More than a time-piece, it is an astronomical and calendrical computer. A celestial globe in front of the main cabinet tracks the positions of five thousand stars, while a device much like an orrery models the motions of the six inner planets. The current phase of the moon is indicated by a rotating globe, half-gilt and half-black.

If you want to know what time it is, the clock offers a choice of answers. A dial mounted on the celestial globe shows sidereal time, as measured by the earth's rotation with respect to the fixed stars. A larger dial on the front of the clock indicates local solar time, which is essentially what a sundial provides; the prick of noon by this measure always comes when the sun is highest overhead. The pointer for local lunar time is similarly synchronized to the height of the moon. (When the solar and lunar pointers coincide, an eclipse is predicted.) Still another dial, with familiar-looking hour and minute hands, shows mean solar time, which averages out the seasonal variations in the earth's orbit to make all days equal in length, exactly twenty-four hours. A second pair of hands on the same dial shows civil time, which in Strasbourg runs thirty minutes ahead of mean solar time. (The city is half a time zone west of the reference meridian for Central European time.)

To count the years there is an inconspicuous four-digit register that anyone from our age of automobiles will instantly recognize as an odometer. Each December 31 at midnight (that's midnight mean solar time, and thus half an hour late by French official time), the counter rolls over to a new year. The transition from 1999 to 2000 went without a hitch.

There's more. A golden wheel nine feet in diameter, marked off into 365 divisions, turns once a year, while Apollo stands at one side to point out today's date. What about leap years? Presto: an extra day magically appears on the wheel when needed. Each daily slot on the calendar wheel is marked with the name of a saint or with some church occasion to be observed that day. Of particular importance, the occasions include Easter and the other "movable feasts" of the ecclesiastical calendar. Calculating the dates of those holidays, and displaying them correctly on the wheel of days, require impressive feats of mechanical trickery.

Wait! There's even more! The clock is inhabited by enough animated figures to open a small theme park. The day of the week is marked by a slow procession of seven Greco-Roman gods in chariots. At noon each day the twelve apostles appear, saluting a figure of Christ, who blesses each in turn and at the end offers benediction to all present. Every half hour a putto overturns a sandglass, and on the quarter hours another strikes a chime. Still more chimes are sounded by figures representing the four ages of mankind, followed by a skeletal Death, who rings the hours. And a mechanical cock crows on cue, flapping its metal wings.

All of this apparatus is housed in a structure of unembarrassed eclecticism, both stylistic and intellectual. The central tower of the clock is topped with a froth of German-baroque frosting, whereas a smaller turret on the left (which houses the weights that drive the clockwork) has been given a more Frenchified treatment. A third tower, on the right, includes a stone spiral staircase that might have been salvaged from an Italian Renaissance belvedere. In the base of the cabinet, two glass panels allowing a view of brass gear trains are a distinctively nineteenth-century element; they look like the windows of an apothecary shop. The paintings and statues are mainly on religious themes—death and resurrection, fall and salvation— but they also include portraits of Urania (Muse of astronomy) and Copernicus. Another painting portrays Jean Baptiste-Sosimé Schwilgué, whose part in this story I shall return to presently.

Programming with Brass

It's all done with gears. Also pinions, worms, snails, arbors; pawls and ratchets; cams and cam followers; cables, levers, bell cranks, and pivots.

The actual timekeeping mechanism—a pendulum and escapement much like those found in other clocks—drives the gear train for mean solar time. All the other astronomical and calendrical functions are derived from this basic, steady motion. For example, local solar time is calculated by applying two corrections to mean solar time. The first correction compensates for seasonal changes in day length, the second for variations in the earth's orbital velocity as it follows its slightly elliptical path around the sun. The corrections are computed by a pair of "profile wheels" whose rims are machined to trace out a graph of the appropriate mathematical function. A roller, following the profile as the wheel turns, adjusts the speed of the local-solar-time pointer accordingly. The computation of lunar motion requires five correction terms and five profile wheels. They all have names: anomaly, evection, variation, annual equation, reduction.

The overall accuracy of the clock can be no better than the adjustment of the pendulum, which requires continual intervention, but for the subsidiary timekeeping functions there is another kind of error to be considered as well. Even if the mean time is exact, will all the solar and lunar and planetary indicators keep pace correctly? The answer depends on how well celestial motions can be approximated by the arithmetic of rational numbers, as embodied in gear ratios. The Strasbourg clock comes impressively close. For example, the true sidereal day is 23 hours, 56 minutes, 4.0905324 seconds, whereas the mean solar day is exactly 24 hours (by definition). The ratio of these intervals is 78,892,313 to 79,108,313, but grinding gears with nearly 80 million teeth is out of the question. The clock approximates the ratio as $1 + (^{450}/_{611} \times ^{1}/_{269})$, which works out to a sidereal day of 23 hours, 56 minutes, 4.0905533 seconds. The error is less than a second per century.

The most intricate calculations are those for leap years and the movable feasts of the church. The rule for leap years states that a year N has an extra day if N is divisible by 4, unless N is also divisible by 100, in which case the year is a common year, with only the usual 365 days; but if N is also divisible by 400, the year becomes a leap year again. Thus 1700, 1800, and 1900 were all common years (at least in those parts of the world that had adopted the Gregorian calendar), but 2000 had a February 29. How can you encode such a nest of if-then-else rules in a gear train? The clock has a wheel with twenty-four teeth and space for an omitted twenty-fifth. This wheel is driven at a rate of one turn per century, so that every four years a tooth comes into position to actuate the leap-year mechanism. The gap where the twenty-fifth tooth would be takes care of the divisible-by-100 exception. For the divisible-by-400 exception to the divisible-by-100 exception, a further adjustment is needed. The key is a second wheel that turns once every 400 years. It carries the missing twenty-fifth tooth and slides it into place on every fourth revolution of the century wheel, just in time to trigger the quadricentennial leap year.

The display of leap years calls for as much ingenuity as their calculation. On the large calendar ring, an open space between December 31 and January 1 bears the legend "Commencement de l'année commune" ("Start of common year"). Shortly before midnight on each December 31 when a leap year is about to begin, a sliding flange that carries the first sixty days of the year ratchets backward by the space of one day, covering up the word "commune" at one end of the flange and at the same time exposing February 29 at the other end. The flange remains in this position throughout the year, then shifts forward again to cover up the 29 and reveal "commune" just as the following year begins.

The rules for finding the date of Easter are even more intricate than the leap-year rule. Donald Knuth, in his *Art of Computer Programming*, remarks: "There are many indications that the sole important application of arithmetic in Europe during the Middle Ages was the calculation of Easter date." Knuth's version of a sixteenth-century algorithm for this calculation has eight major steps, some of which are fairly complex. Here's step five:

> Set $E \leftarrow (11G + 20 + Z - X)$ mod 30. If $E = 25$ and the golden number G is greater than 11, or if $E = 24$, then increase E by 1. (E is the so-called "epact," which specifies when a full moon occurs.)

Programming a modern computer to perform the Easter calculation requires some care; programming a box of brass gears to do the arithmetic is truly a tour de force. I have stared at diagrams of the gears and linkages and tried to trace out their action, but I still don't fully understand how it all fits together.

In the abstract, it's not too hard to see how a mechanical linkage could carry out the basic steps of the epact calculation given above. A wheel with 30 teeth or cogs would ratchet $11G$ notches clockwise, then it would add 20 steps more in the same direction, then another Z steps; finally it would turn X steps counterclockwise. The "mod 30" part of the program—reducing the sum modulo 30 (so that 30 becomes 0, 31 becomes 1, and so forth)—would be taken care of automatically by doing the arithmetic on a circle with 30 divisions. So far so good. The 30-tooth wheel does exist in the Strasbourg clock, and it is even helpfully labeled "Epacte." Where I get lost is in trying to understand the various lever arms and rack-and-pinion assemblies that drive the epact wheel, and the cam followers that communicate its state to the rest of the system. There appear to be a number of optimizations in the gear works, which doubtless save a little brass but make the operation more obscure. Perhaps if I had a model I could take apart and put together again . . .

But never mind my failures of spatiotemporal reasoning. The mechanism *does* work. Each New Year's Eve a metal tag that marks the date of Easter slides along the circumference of the calendar ring and takes up a position over the correct Sunday for the coming year. (The date of Easter can range from March 22 to April 25.) All the other movable feasts of the church are a fixed number of days before or after Easter, so the indicators of their dates are rigidly linked to the Easter tag and move along with it.

Making It Go

The present Strasbourg clock is the third in a series. The first was built in the middle of the fourteenth century, just as the cathedral itself was being completed with the addition of a spire that made it the tallest structure in Europe. That original clock had animated figures of the three Magi who bowed down before the Virgin and Child every hour on the hour. Little else is known of it, and all that survives is a mechanical rooster, ancestor of the current cock of the clock.

By the middle of the sixteenth century, the Clock of the Three Kings was no longer running and no longer at the leading edge of horological technology. To supervise an upgrade, the Strasbourgeois hired Conrad Dasypodius, the professor of mathematics at Strasbourg, as well as the clockmaker Isaac Habrecht and the artist Tobias Stimmer. These three laid out the basic plan of the instrument still seen today, including the three-turreted case and most of the paintings and sculptures. A curiosity surviving from this era is the portrait of Copernicus— a curiosity because the planetary display on the Dasypodius clock portrayed not the sun-centered Copernican system but the earth-centered Ptolemaic one. The second clock lasted another two hundred years, give or take.

The story of the third clock starts with an anecdote so charming that I can't bear to look too closely into its authenticity. Early in the nineteenth century, the story goes, a beadle was giving a tour of the cathedral, and mentioned that the clock had been stopped for twenty years and no one knew how to fix it. A small voice piped up: "I will make it go!" The boy who made this declaration was Jean Baptiste-Sosimé Schwilgué, who made good on his promise forty years later.

There was mild conflict over the terms of Schwilgué's commission. He wanted to build a wholly new clock; the cathedral administration wanted to repair the old one. They compromised: he gutted the works, but kept the case, and built his new indicators and automata to fit the old design. The new mechanism was first started up on October 2, 1842.

Schwilgué was clearly thinking long-term when he undertook the project. As I have already noted, the leap-year mechanism includes components that engage only once every four hundred years—parts that were tested for the first time in 2000 and will lie dormant again until 2400. Such very rare events might have been left for manual correction. It would have been only a small imposition on the clock's maintainers to ask that the hands be reset every four centuries. But Schwilgué evidently took pride and pleasure in getting the details right. He couldn't know if the clock would still be running in 2000 or 2400, but he could build it in such a way that if it *did* survive, it would not perpetrate error.

The contrast with recent practice in computer hardware and software could hardly be more stark. Many computer systems—even those that survived the Y2K scare—are explicitly limited to dates between 1901 and 2099. The reason for choosing this particular span is that it makes the leap-year rule extremely simple: it's just a test of divisibility by four. Under the circumstances, this design choice seems pretty wimpy. If Schwilgué could take the trouble to fabricate wheels that make one revolution every one hundred and four hundred years, surely a programmer could write the extra line of code needed to check for the century exceptions. The line might never be needed, but there's the satisfaction of knowing it's there.

Other parts of Schwilgué's clock look even further into the future. There is a gear deep in the works of the ecclesiastical computer that turns once every 2,500 years. And the celestial sphere out in front of the clock has a still-slower motion. In addition to the sphere's daily rotation, it pirouettes slowly on another axis to reflect the precession of the equinoxes of the earth's orbit through the constellations of the zodiac. In the real solar system, this stately motion is what has lately brought us to the dawning of the age of Aquarius. In the clock, the once-per-sidereal-day spinning of the globe is geared down at a ratio of 9,451,512 to 1, so that the equinoxes will complete one full precessional cycle after the passage of 25,806 years. (The actual period is now thought to be 25,784 years.) At that point we'll be back to the cusp of Aquarius again, and no doubt paisley bell-bottoms will be back in fashion.

Easter in 11842 Falls on April 3

The odometer of years on the face of the Strasbourg clock, as mentioned above, runs up to 9999. According to some accounts, Schwilgué suggested that if the clock is still going when the counter rolls over to 0000, a numeral 1 could be painted on the case to the left of the thousands digit. The simplicity of this solution suggests that the Y10K crisis may turn out to be even less disruptive than the Y2K one was. After all, appending a digit to the tally is easier than changing one.

Is there any chance the Strasbourg clock will actually run for ten thousand years? No products of human artifice have yet lasted so long, with the exception of cave paintings and some sharpened flints. Stonehenge and the pyramids of Egypt are half that age. The two earlier Strasbourg clocks, built with technology similar to that of the current instrument, both failed after roughly two centuries. Complex machines with moving parts seldom seem to last more than a few hundred years, even with conscientious maintenance. Of course such machines were great rarities until a few hundred years ago, so the age distribution is highly skewed. One might equally well argue that electronic computers cannot work for more than fifty or sixty years, since all the functioning ones are younger than that. Still, the actuarial life expectancy of either the clock or the computer can surely be expressed in three digits or less.

In principle, a machine can last forever if you keep replacing parts as fast as they wear out. For this strategy to succeed, however, not only artifacts but also institutions must survive. Someone must be there to wind the clock and lube it and dust it, day after day. The perils to long-term continuity should be abundantly clear in a border town like Strasbourg, which has been batted back and forth between France and Germany like the child of a contested divorce. Bishops and burghers once fought for control of the city; the cathedral has passed through the hands of Catholics, Protestants, and revolutionary atheists. And yet the stones still stand, and so do the institutions. A single organization, the Oeuvre Notre-Dame, has maintained the cathedral since the thirteenth century.

Even if the clock keeps ticking, however, will anyone in 11842 want to know the date of Easter? For that matter, will people then still be counting the years of the Common Era? No system of timekeeping has endured anywhere near ten millennia. The Roman calendar was abandoned after fifteen hundred years; the Mayan one may have lasted as long as two thousand years, the Egyptian possibly three thousand. According to the Hebrew calendar, the tally of years is now well past fifty-seven hundred, but that's not to say that anyone has been faithfully marking the days and months since 1 Tishri 1. Meanwhile, other calendars have come and gone. If Schwilgué had rebuilt the Strasbourg clock just a few decades earlier, it would have listed dates in Brumaire, Thermidor, Fructidor, and the other months decreed by the French Revolution, and the year would now be in the low 200s.

The Long Now

I want to address another question. Even if a clock can be kept in working order, and even if the calendar it keeps retains some meaning, is the building of such multimillennial machines a good idea? I have my doubts, and they have been redoubled by a recent proposal to build another ten-thousand-year clock.

The new plan comes from Danny Hillis, the architect of the Connection Machine, an innovative and widely admired supercomputer of the 1980s. Another of Hillis's projects was a computer made entirely of Tinkertoys, which has a distant familial connection with Schwilgué's ecclesiastical computer. Together with several friends and colleagues, Hillis has proposed building a clock described as "the world's slowest computer," whose function is just to keep going as long as possible. The project is outlined in *The Clock of the Long Now*, a book by Stewart Brand, the instigator of *The Whole Earth Catalog*.

Technical details of the Long Now clock remain to be worked out, but the provisional design that Brand describes has a torsion pendulum (one that twists rather than swings) and a digital counter of pendulum oscillations instead of an analog gear train. Although the counter is digital, it is emphatically *not* electronic; Hillis's design uses mechanical wheels and pegs to count in binary notation from zero up to some predefined constant, such as the number of seconds in a year.

The plan is to build several clocks, of increasing grandeur. A prototype will be eight feet high. A twenty-foot model will be placed in a large city for ease of access, and then a sixty-footer will be installed somewhere out in the desert for safekeeping. Here is one of Hillis's visions of how the full-size clock might be experienced:

> Imagine the clock is a series of rooms. In the first chamber is a large, slow pendulum. This is your heart beating, but slower. In the next chamber is a simple twenty-four-hour clock that goes around once a day. In the next chamber, just a Moon globe, showing the phase of the lunar month. In the next chamber is an armillary sphere tracking the equinoxes, the solstices, and the inclination of the Sun . . . The next chamber is the Lifetime room—a single blank, featureless disk of soft stone that rotates once a lifetime, onto which you can carve your own mark.
>
> The final chamber is much larger than the rest. This is the calendar room. It contains a ring that rotates once a century and the 10,000-year segment of a much larger ring that rotates once every precession of the equinoxes. These two rings intersect to show the current calendrical date.

The motive for building this monument to slow motion is not timekeeping per se; Hillis is not worried about losing count of the centuries. The aim is psychological. The clock is meant to encourage long-term thinking, to remind people of the needs and claims of future generations. The preamble to the project summary begins: "Civilization is revving itself into a pathologically short attention span. The trend might be coming from the acceleration of technology, the short-horizon perspective of market-driven economics, the next-election perspective of democracies, or the distractions of personal multitasking." The big slow clock would offer a counterpoise to these frenetic tendencies; it would "embody deep time."

The wisdom of planning ahead, husbanding resources, saving something for those who will come after, leaving the world a better place—it's hard to quibble with all that. Concern for the welfare of one's children and grandchildren is surely a virtue—or at least a Darwinian imperative—and more-general benevolence toward future inhabitants of the planet is also widely esteemed. But if looking ahead two or three generations is good, does that mean looking ahead twenty or thirty generations is better? What about two hundred or three hundred generations? Perhaps the answer depends on how far ahead you can actually see.

The Long Now group urges us to act in the best interests of posterity, but beyond a century or two I have no idea what those interests might be. To assume that the values of our own age embody eternal verities and virtues is foolish and arrogant. For all I know, some future generation will thank us for burning up all that noxious petroleum and curse us for exterminating the smallpox virus.

From a reading of Brand's book, I don't sense that the Long Now organizers can see any further ahead than the rest of us; as a matter of fact, they seem to be living in quite a short Now. All those afflictions listed in their preamble—the focus on quarterly earnings and quadrennial elections and so forth—are bugaboos of recent years and decades. They would have been incomprehensible a few centuries ago, and there's not much reason to suppose they will make anybody's list of pressing concerns a few centuries hence, much less in ten thousand years.

The emphasis on the superiority of binary digital computing is something else that puts a late-twentieth-century date stamp on the project. A time may come when Hillis's binary counters will look just as quaint as Schwilgué's brass gears.

Long-term thinking is really hard. Of course that's the point of the Long Now project, but it's also a point of weakness. It's hard to keep in mind that what seems most steadfast over the human life span may be evanescent on a geological or astronomical timescale. Consider the plan to put one clock in a city (New York, say) and another in a desert (Nevada). This makes sense now, but will New York remain urban and Nevada unpopulated over the next ten thousand years? Many a desolate spot in the desert was once a city, and vice versa. (On the other hand, maybe Nevada isn't such a bad choice. They could build the clock in Yucca Mountain, the proposed repository of one product of civilization we can count on lasting ten thousand years: the radioactive wastes from nuclear power generation.)

Needless to say, the difficulty of predicting the future is no warrant to ignore it. The Y2K scare offered clear evidence that a time horizon of two digits is too short. But four digits is plenty. If we take up the habit of building machines meant to last past 10000, or if we write our computer programs with room for five-digit years, we are not doing the future a favor. We're merely nourishing our own delusions.

Chronocolonialism

In the sixteenth century, Dasypodius and his colleagues could have chosen to restore the two-hundred-year-old Clock of the Three Kings in Strasbourg Cathedral, but instead they ripped out all traces of it and built a new and better clock. A bit more than two hundred years later, Schwilgué was asked to repair the Dasypodius clock, but instead he eviscerated it and installed his own mechanism in the hollowed-out carcass. He built a new and better clock, good for ten thousand years. Today, after another two centuries, the Long Now group is not threatening to destroy the Schwilgué clock, but neither are they working to ensure its longevity. They ignore it. They want to build a newer, better, different clock, good for ten thousand years.

I begin to detect a pattern. The fact is, winding and dusting and fixing somebody else's old clock is boring. Building a brand-new clock of your own is much more fun, especially if you can pretend that it's going to inspire awe and wonder for ages to come. So why not have the fun now and let the next three hundred generations do the boring bit?

If I thought that Hillis and his associates might possibly succeed in this act of chronocolonialism—enslaving future generations to maintain our legacy systems—I would consider it my own duty to posterity to oppose the project, even to sabotage it. But in fact I don't worry. I have faith in the future. Sometime in the 2200s a small child touring the ruins of the Clock of the Long Now will proclaim, "*I will make it go!*" And that child will surely scrap the whole mess and build a new and better clock, good for ten thousand years.

AFTERTHOUGHTS

Of all the essays collected in this volume, this meditation on the theme of things-built-to-last turned out to be the most perishable. It was written for a particular occasion—the end of the second millennium of the Common Era—and was published in the November–December 1999 issue of *The Sciences*. When I began to prepare the essay for re-publication here, I found that much of the discussion was closely tied to the preoccupations of that single moment, especially the impending "Y2K crisis." Obviously, I wasn't thinking long-term when I wrote the piece.

I have done some light editing to liberate the essay from its captivity in 1999, without trying to rewrite it entirely from the perspective of 2007. (This moment, too, shall pass, after all.) I have also taken the opportunity to restore a few passages that had to be omitted from the original magazine version for lack of space. Finally, I have corrected an embarrassing error.

Here's the story of the error. I had believed that Schwilgué's simple ploy of painting a 1 next to the year counter would not fully adapt the clock to the post-10000 era, because the Easter calculation would be incorrect. The algorithm for calculating the date of Easter takes the year number as its input and produces a month and a date as its output. I had written a little computer program based on this algorithm, which seemed to indicate that the clock's calculation would go awry after 9999. For example, the date of Easter in 11999 (if anyone is paying attention then) will be April 11; if the ecclesiastical computer inside the clock were to see only the final four digits of this number, it would calculate the date of Easter for 1999, which was April 4. Thus the clock would forever repeat a ten-thousand-year cycle of Easter dates.

The flaw in my reasoning was to assume that the gear train inside the clock worked the same way as my program. In actuality, the ecclesiastical computer is oblivious to the numerical value of the year; it performs its calculation progressively, year by year, in effect simulating all the motions of the sun and the moon that ultimately determine the Easter date. The year number never enters into the calculation; the church calendar would continue to be updated correctly even if the counter of years were to stop.

What's embarrassing about this error is that my faulty understanding was pointed out even before publication, by Peter Brown, who was then the editor of *The Sciences*; but I was sure I was right. It was a letter to the editor from Bob Conley of Anchorage, Alaska, that finally convinced me.

There is another small error; I have allowed it to stand in the text, but I should acknowledge it here. When I wrote that cave paintings and stone tools are the only man-made artifacts to have survived ten thousand years or more, I was forgetting about pottery. The jomon pottery of Japan is thought to go back at least eleven thousand years and perhaps substantially longer.

What has become of the Clock of the Long Now? A first prototype was completed just in time to bong twice at midnight on December 31, 2000. That clock is now on display at the Science Museum in London. A second prototype is under construction in California, and the Long Now Foundation has purchased a site in eastern Nevada—not too far from Yucca Mountain!—for the big stone clock. For more information and progress reports, see www.longnow.org.

The debate over global warming, which has heated up considerably since this essay first appeared, prompts a further comment on the need for long-term thinking, and on the difficulty of seeing into the distant future. The idea that actions we take now—or fail to take—might alter the earth's climate for thousands of years to come argues that we cannot shirk responsibility for the fate of the planet. Whether or not we bother to give any thought to the future, we will determine its shape.

But the prospect of a major shift in climate is also a reminder that the future is a different world—different from the one we inhabit now, and also probably different from any that we can imagine. The current fear is that Florida will become a shallow sea, and the Corn Belt will be growing nothing but cactus. It could happen. But so could the opposite: the glaciers might descend again and scrape the continent clean as far south as St. Louis. For optimists, there's also the possibility that by the year 12000 people may have acquired the means and the wisdom to control the planet's wilder divagations.

When I was younger, the peril to human survival that seemed most worrisome was a nuclear shoot-out that could leave the earth uninhabitable. That threat has not disappeared, although it has retreated from public consciousness. There have also been periods when overpopulation, pollution, and the exhaustion of resources appeared to be the principal agents of doom. Now the focus is on climate. In reciting this history of shifting hazards, I don't mean to belittle any of them; they are all to be taken seriously. I simply want to point out that when we make an earnest effort to think about "the long run," our vision of the distant future always seems to reflect mainly the concerns of the present moment.

Trying to make life a little better for the great-great- . . . great-grandkids is a worthy goal; conversely, policies that cater solely to our own comfort and convenience, ignoring the welfare of future generations, are reprehensible. If building the Clock of the Long Now will help to remind people that life goes on—or that we *hope* it goes on—then there's surely no harm in it. But keep in mind that it's an exercise for the benefit of those who build it, not for those to whom we supposedly bequeath it.

CHAPTER 2

Random Resources

Randomness is not something we usually look upon as a vital natural resource, to be carefully conserved lest our grandchildren run short of it. On the contrary, as a close relative of chaos, randomness seems to be all too abundant and ever present. Everyone has a closet or a file drawer that offers an inexhaustible supply of disorder. Entropy—another cousin of randomness—even has a law of nature saying it can only increase. And, anyway, even if we were somehow to use up all the world's randomness, who would lament the loss? Fretting about a dearth of randomness seems like worrying that humanity might use up its last reserves of ignorance.

Nevertheless, there is a case to be made for the proposition that high-quality randomness is a valuable commodity. Many events and processes in the modern world depend on a steady supply of the stuff. Furthermore, we don't know how to manufacture randomness; we can only mine it from those regions of the universe that have the richest deposits, or else farm it from seeds gathered in the natural world. So, even if we have not yet reached the point of clear-cutting the last proud acre of old-

growth randomness, maybe it's not too early to consider the
question of long-term supply.

The Randomness Industry

To appreciate the value of randomness, imagine a world without it. What would replace the referee's coin flip at the start of
a football game? How would a political poll taker select an un-
biased sample of the electorate? Then of course there's the Las
Vegas problem. Slot machines devour even more randomness
than they do silver dollars. Inside each machine an electronic
device spews out random numbers twenty-four hours a day,
whether or not anyone is playing.

There's also a Monte Carlo problem. I speak not of the
Mediterranean principality but of a computer simulation technique named for that place. The Monte Carlo method got its
start in the 1940s at Los Alamos, where physicists designing
nuclear weapons were struggling to predict the fate of neutrons
moving through uranium and other materials. Neutrons are
the initiators of the nuclear chain reaction. On striking an
atomic nucleus, a neutron might bounce off in a new direction,
or it might be absorbed; in the latter case, there's a further
chance the absorbed neutron might cause the nucleus to break
apart. This splitting process (fission) emits more neutrons,
which can go on to induce still more fission events. The crucial
question is whether the supply of available neutrons is growing
or shrinking. To find an answer, the Los Alamos group decided to trace thousands of neutron paths by computer simulation. Whenever a neutron encountered a nucleus, a random
number determined the outcome of the event—reflection, ab-
sorption, or fission. The idea of enlisting randomness in the
cause of science was a novelty then; it may even have seemed
a trifle naughty. Today, however, the Monte Carlo method is a

major industry not only in physics but also in economics and some areas of the life sciences, not to mention hundreds of rotisserie baseball leagues.

Many computer networks would be deadlocked without access to randomness. When two nodes on a network try to

Four specimens of randomness employ black and white dots to represent the binary digits 1 and 0. The four sources of random bits are a series of coin flips, a table of random numbers published by the RAND Corporation, a defunct Web site that extracted random numbers from the moving blobs inside Lava Lites, and the Pennsylvania lottery.

speak at once, neither can be heard, and politeness is not always enough to break the impasse. Suppose each computer were programmed to wait a certain fixed interval and then to try again; if all computers followed the same rule, they'd keep knocking heads repeatedly until the lights went out. The networking protocol called Ethernet solves this problem by deliberately not giving a fixed rule. Instead, each machine picks a random number x in the range between 1 and some constant n; the machine then waits x units of time before retransmitting. The probability of a second collision is reduced to $1/n$. This idea was first implemented in a Hawaiian "packet radio" network called ALOHAnet. When it was later adopted for the Ethernet protocol, the reliance on randomness was viewed with a certain skepticism. But today most networked computers are connected via Ethernet, and their occasional random rounds of "After you . . . ," "No, after *you* . . ." pass quite unnoticed.

Computer science has a whole technology of "randomized algorithms." On first acquaintance the very idea of a randomized algorithm may seem slightly peculiar: an algorithm is supposed to be a deterministic procedure—one that allows no scope for arbitrary choice or caprice—so how can it be randomized? The contradiction is resolved by making the randomness a resource external to the algorithm itself. Where an ordinary algorithm is a black box receiving a stream of bits as input and producing another stream of bits as output, a randomized algorithm has a second input stream made up of random bits.

Sometimes the advantage of a randomized algorithm is clearest when you take an adversarial view of the world. Randomness is what you need to foil an adversary who wants to guess your intentions or predict your behavior. Suppose you are writing a program to search a list of items for some specified target. Given any predetermined search strategy—left to right, right to left, middle outward—an adversary can arrange the list so that the target item is always in the last place you look. But a randomized version of the procedure can't be outguessed so easily; the adversary can't know where to hide the target, because the program doesn't decide where to search until it begins reading random bits. In spite of the adversary's best efforts, you can expect to find the target after sifting through about half the list.

Hiding Signals in the Noise

Still another consumer of randomness is cryptography, where calculated disorder is the secret to secrecy. The value of randomness is obvious in one of the simplest cipher systems, which also happens to be the strongest of them all. During World War I, Gilbert S. Vernam, an engineer with AT&T, invented a scheme for secret communication based on a printing telegraph, a machine driven by a punched paper tape. Vernam's enciphering machine combines two such tapes, one punched with the "plaintext" message and the other, called the key tape, bearing a random pattern of the same length. The machine treats the patterns of punched holes as binary numbers and combines the individual bits by addition modulo 2, so that $0+1=1$ and $1+0=1$ but $0+0=0$ and $1+1=0$. The result of all the additions is transmitted as the secret "ciphertext." Because the addition process is its own inverse, the recipient at the other end of the telegraph line can use an identical machine to combine the ciphertext with the same random key tape to recover the plaintext.

Claude E. Shannon, also of AT&T, later proved that Vernam's cipher is absolutely secure. That is, if the bits of the key are truly random, and if they are used only once, an eavesdropper who intercepts the encrypted message can learn nothing from it about the plaintext, no matter how much time and effort and computational horsepower are brought to bear on the problem. Even trying every possible key is of no avail; since every key is equally likely, so is every plaintext message. Shannon also showed that no cipher with a key shorter than the message could offer the same degree of security.

The long random key is the strength of the Vernam cipher but also its weakness. Before two parties can establish secret communication, they must make and exchange identical copies of a random key as long as all the messages they plan to send. Where is all that randomness to be found? (Vernam suggested "working the keyboard at random"—a practice that would horrify modern code clerks.) Because of this considerable inconvenience, Vernam ciphers are used only for the hush-hushest channels, such as the Washington-Moscow hotline.

Much of the emphasis in recent cryptological research has been on ways to get by with less randomness, but a recent proposal takes a step in the other direction. The idea is to drown an adversary in a deluge of random bits. The first version of the scheme was put forward in 1992 by Ueli M. Maurer of the Swiss Federal Institute of Technology; more recent refinements have come from Michael O. Rabin of Harvard University and his student Yan Zong Ding (now at Georgia Tech).

The heart of the plan is to set up a public beacon—perhaps a satellite—continually spewing out random bits at a rate so prodigious that no one could store more than a small fraction of them. Parties who want to communicate in privacy share a relatively short key that they both use to select a sequence of random bits from the public broadcast; the selected bits serve as an enciphering key for their messages. An eavesdropper cannot decrypt an intercepted message without knowing which segment of the random broadcast to use as the key. The brute-force strategy of trying all possible segments is foiled, again, by the impracticality of storing all that nonsense.

How much randomness would the beacon have to broadcast? Rabin and Ding mention a rate of fifty gigabits per second, which would fill up some 800,000 CDs per day.

Supply-Side Issues

Whatever the purpose of randomness, and however light or heavy the demand, it seems like producing the stuff ought to be a cinch. At the very least it should be easier to make random bits than nonrandom ones, in the same way that it's easier to make a mess than it is to tidy up. If computers can perform long and intricate calculations where a single error could spoil the entire result, then surely they should be able to churn out some patternless digital junk. But they can't. There is no computer program for randomness.

Of course most computer programming languages will cheerfully offer to generate random numbers for you. In Lisp the expression "(random 100)" produces an integer in the range between 0 and 99, with each of the 100 possible values having equal probability. But these are *pseudo*random numbers: they "look" random, but under the surface there is nothing unpredictable about them. Each number in the series depends on those that went before. You may not immediately perceive the rule in a series like 58, 23, 0, 79, 48. . . , but it's just as deterministic as 1, 2, 3, 4. . .

The only source of true randomness in a sequence of pseudorandom numbers is a "seed" value that gets the series started. If you supply identical seeds, you get identical sequences; different seeds produce different sets of numbers. The crucial role of the seed was made clear in the 1980s by Manuel Blum, now of Carnegie Mellon University. He pointed out that a pseudorandom generator does not actually generate any randomness; it stretches or dilutes whatever randomness is present in the seed, spreading it out over a longer series of numbers like a drop of pigment mixed into a gallon of paint.

For most purposes, pseudorandom numbers serve perfectly well. Almost all Monte Carlo work is based on them. Even for some cryptographic applications—where standards are higher and unpredictability is everything—Blum and others have invented pseudorandom generators that meet most needs. Nevertheless, true randomness is still in demand, if only to supply seeds for pseudorandom generators. And if true randomness cannot be created in any mathematical operation, then it will have to come from some physical process.

Extracting randomness from the material world also sounds like an easy enough job. Unpredictable events are all around us: the stock market tomorrow, the weather next week, the orbital position of Pluto in fifty million years. Yet finding events that are totally patternless turns out to be quite difficult. The stories of the pioneering seekers after randomness are chronicles of travail and disappointment.

Consider the experience of the British biometrician W.F.R. Weldon and his wife, the former Florence Tebb. Evidently, they spent many an evening rolling dice together—not for money or sport but for science, collecting data for a classroom demonstration of the laws of probability. They did their best, but in 1900 Karl Pearson analyzed 26,306 of the Weldons' dice throws and found deviations from the expected distribution; there was an excess of fives and sixes.

In 1901 William Thomson, Lord Kelvin, tried to carry out what would now be called a Monte Carlo experiment, but he ran into trouble generating random numbers. In a footnote he wrote: "I had tried numbered billets (small squares of paper) drawn from a bowl, but found this very unsatisfactory. The best mixing we could make in the bowl seemed to be quite insufficient to secure equal chances for all the billets."

In 1925 L.H.C. Tippett had the same problem. When he tried to make a random selection from a thousand cards in a bag, "it was concluded that the mixing between each draw had not been sufficient, and there was a tendency for neighbouring draws to be alike." Tippett devised a more elaborate randomizing procedure, and two years later he published a table of 41,600 random digits. But in 1938 G. Udny Yule submitted Tippett's numbers to statistical scrutiny and reported evidence of "patchiness."

Ronald A. Fisher and Frank Yates compiled another table of 15,000 random digits, using two decks of playing cards to select numbers from a large table of logarithms. When they were done, they discovered an excess of sixes, and so they replaced fifty of them with other digits "selected at random." (Two of their statistical colleagues, Maurice G. Kendall and Bernard Babington Smith, comment mildly: "A procedure of this kind may cause others, as it did us, some misgiving.")

Hard-Wired Randomness

The ultimate random-number table arrived with a thump in 1955, when the RAND Corporation published a six-hundred-page tome titled *A Million Random Digits with 100,000 Normal Deviates*. The RAND randomizers used an "electronic roulette wheel" that selected one digit per second. Despite the care taken in the construction of this device, "production from the original machine showed statistically significant biases, and the engineers had to make several modifications and refinements of the circuits." Even after this tune-up, the results of the monthlong run were still unsatisfactory; RAND had to reshuffle the numbers before the tables passed statistical tests.

Today there is little interest in publishing tables of numbers, but machines for generating randomness are still being built.

Many of them find their source of disorder in the thermal fluctuations of electrons wandering through a resistor or a semi-
conductor junction. This noisy signal is the hiss or whoosh you
hear when you turn up an audio amplifier's volume control.
Traced by an oscilloscope, it certainly looks random and unpredictable, but converting it into a stream of random bits or
numbers is not straightforward.

The obvious scheme for digitizing noise is to measure the
signal at certain instants and emit a 1 if the voltage is positive
or a 0 if it is negative. But it's hard to build a measuring circuit
with a precise and consistent threshold between positive and
negative voltage. As components age, the threshold drifts, causing a bias in the balance between 1s and 0s. There are circuits
and computational tricks to correct this problem, but the need
for such fixes suggests just how messy it can be getting a physical device to conform to a mathematical ideal—even when the
ideal is that of pure messiness.

Another popular source of randomness is the radioactive
decay of atomic nuclei, a quantum phenomenon that seems to
be near the ultimate in unpredictability. A simple random-
number generator based on this effect might work as follows.
A Geiger-Müller tube (the device at the heart of a Geiger
counter) detects a decay event, while in the background a free-
running oscillator generates a high-frequency square-wave signal—a train of positive and negative electrical pulses. At the
instant of a nuclear decay, the square wave is sampled, and a binary 1 or 0 is output according to the polarity of the pulse at
that moment. Again there are engineering pitfalls. For example, the circuitry's "dead time" after each event may block detection of closely spaced decays. And if the positive and negative
pulses in the square wave differ in length even slightly, the output will be biased in favor of one digit or the other.

Hardware random-number generators are available as off-the-shelf components you can plug into a port of your computer. Most of them rely on thermal electronic noise. If your computer has one of the later Intel processors, you don't need to plug in a peripheral: the random-number generator is built into the CPU chip. Various Web sites serve up free samples of randomness, but like so much on the Web, the sites tend to come and go. Perhaps the most famous was the LavaRand service of Silicon Graphics, Inc., where random bits were extracted from images of the erupting blobs inside six Lava Lite lamps. Sadly, the lamps have gone out. A successor site, LavaRnd.org, still produces randomness on demand, but without the head shop hardware. Another Web site, known as HotBits, produces random bits from radioactive decay events.

The Empyrean and the Empirical

As a practical matter, reserves of randomness certainly appear adequate to meet current needs. Consumers of random bits need not fear rolling blackouts in steamy weather. But what of the future? The great beacon of randomness proposed by Rabin and Ding would require technology that remains to be demonstrated. They envision broadcasting fifty billion random bits per second, but randomness generators today typically run at speeds closer to fifty *thousand* bits per second.

The prospect of scaling up by a factor of a million demands attention to quality as well as quantity. For most commodities, quantity and quality have an inverse relation. A laboratory buying milligrams of a reagent may demand 99.9 percent purity, whereas a factory using carloads can tolerate a lower standard. In the case of randomness, the trade-off is turned upside down. If you need just a few random numbers, any source will do; it's hard to spot biases in a handful of bits. But a Monte Carlo experiment burning up billions of random numbers is exquisitely sensitive to the faintest trends and patterns. The more randomness you consume, the better it has to be.

Why is it hard to make randomness? The fact that maintaining perfect order is difficult surprises no one; but it comes as something of a revelation that perfect disorder is also beyond our reach. As a matter of fact, perfect disorder is the more troubling concept—it is hard not only to attain but also to define or even to imagine.

A widely known definition of randomness was formulated in the 1960s by Gregory J. Chaitin of IBM and by the Russian mathematician A. N. Kolmogorov. The definition says that a sequence of bits is random if the shortest computer program for generating the sequence is at least as long as the sequence itself. The binary string 101010101010 is not random, because there is an easy rule for creating it, whereas 111010001011 is unlikely to have a generating program much shorter than "print 111010001011." It turns out that almost all strings of bits are random by this criterion—they have no concise description—and yet no one has ever exhibited a single string that is certified to be random. The reason is simple: the first string certified to have no concise description would thereby acquire a concise description—namely that it's the first such string.

The Chaitin-Kolmogorov definition is not the only aspect of randomness verging on the paradoxical or the ironic. Here is another example: true random numbers, captured in the wild, are clearly superior to those bred in captivity by pseudorandom generators—or at least that's what the theory of randomness implies. But George Marsaglia of Florida State University has run the output of various hardware and software generators through a series of statistical tests. The best of the pseudorandom generators earned excellent grades, but three hardware devices flunked. In other words, the fakes look more convincingly random than the real thing.

To me the strangest aspect of randomness is its role as a link between the world of mathematical abstraction and the universe of ponderable matter and energy. The fact that randomness requires a physical rather than a mathematical source is noted by almost everyone who writes on the subject, and yet the oddity of this situation is not much remarked.

Mathematics and theoretical computer science inhabit a realm of idealized and immaterial objects: points and lines, sets, numbers, algorithms, Turing machines. For the most part, this world is self-contained; anything you need in it, you can make in it. If a calculation calls for the millionth prime number or the cube root of 2, you can set the computational machinery in motion without ever leaving the precincts of mathland. The one exception is randomness. When a calculation asks for a random number, no mathematical apparatus can supply it. There is no alternative but to reach outside the mathematical empyrean into the grubby world of noisy circuits and decaying atomic nuclei. What a strange maneuver! If some purely mathematical statement—say the formula for solving a quadratic equation—depended on the mass of the earth or the diameter of the hydrogen atom, we would find this disturbing or absurd. Importing physical randomness into mathematics crosses the same boundary.

Of course there is another point of view: if we choose to look upon mathematics as a science limited to deterministic operations, it's hardly a surprise that absence of determinism can't be found there. Perhaps what is really extraordinary is not that randomness lies outside mathematics but that it exists anywhere at all.

Or does it? The savants of the eighteenth century didn't think so. In their clockwork universe the chain of cause and effect was never broken. Events that appeared to be random were merely too complicated to submit to a full analysis. If we failed to predict the exact motion of an object—a roving comet, a spinning coin—the fault lay not in the unruliness of the movement but in our ignorance of the laws of physics or the initial conditions.

The issue is seen differently today. Quantum mechanics has cast a deep shadow over causality, at least in microscopic domains. And "deterministic chaos" has added its own penumbra, obscuring the details of events that might be predicted in principle, but only if we could gather an unbounded amount of information about them. To a modern sensibility, randomness reflects not just the limits of human knowledge but some inherent property of the world we live in. Nevertheless, it seems fair to say that most of what goes on in our neighborhood of the universe is deterministic. Coins spinning in the air and dice tumbling on the felt are not conspicuously quantum-mechanical or chaotic systems. We choose to describe their behavior through the laws of probability only as a matter of convenience; there's no question the laws of angular momentum are at work behind the scenes. If there is any genuine randomness to be found in such events, it is the merest sliver of quantum uncertainty. Perhaps this helps to explain why digging for randomness in the flinty soil of physics is such hard work.

AFTERTHOUGHTS

In the half-dozen years since this essay first appeared in *American Scientist*, there have been no news reports of acute shortages of randomness—no riots, rationing, or price controls. So, was it all a false alarm? Clearly, there's no reason for panic. If there's a randomness crisis, it's not one that's going to imperil civilization. But generating random bits in sufficient quantity and quality is still an important practical challenge.

My statement that randomness does not exist in the realm of pure mathematics was challenged by readers who pointed to numbers such as π and e, whose digits form famously random sequences. These numbers (along with many others) are considered "normal numbers," a slightly confusing term that has nothing to do with the normal distribution of statistics but rather describes a number in which all digits appear with the same probability, and likewise all pairs of digits, triplets of digits, and so on. In other words, when a normal number is written in base 10, a tenth of the digits are 0s, a tenth are 1s, a tenth are 2s, and so forth. For pairs of digits such as 00, 01, 02, and 03, the probability in all cases is $\frac{1}{100}$, since there are 100 possible pairs. The status of normal numbers in mathematics is perplexing. On the one hand, there's an ironclad proof that "almost all" numbers are normal; if you throw a dart at the real number line, you are almost certain to land on a normal number. On the other hand, when it comes to specific numbers, only a few examples have been proved normal. For numbers such as π and e, the situation is murky: everyone believes they are normal, and thus that they offer untapped reserves of randomness, but no one can prove their normality.

Shu-Ju Tu and Ephraim Fischbach of Purdue University have looked into using the digits of π as a practical source of random numbers, subjecting the digit stream to the same kinds of statistical tests that have been developed for evaluating pseudorandom numbers. They find that π makes a mediocre random-number generator.

The subject of randomness seems to bring out a philosophical streak in many of us. Roger S. Pinkham of the Stevens Institute of Technology challenged me to reexamine my own philosophical assumptions: "What makes you think emissions of alpha particles, electronic noise, and other 'real world' phenomena are random?" It's a good question. After all, in many cases randomness is just a convenient fiction. We treat the outcome of a coin toss as random because that's easier than trying to account for all the forces acting on the coin—the initial impulse, the pull of gravity, all the subtle effects of aerodynamics. But with enough analytic effort, there would be no mystery in the motion of the coin. Persi Diaconis, a statistician at Stanford University and also an accomplished magician, has taught himself to flip a coin so consistently that he can produce the same outcome ten times in a row. (He can also shuffle a deck of cards in a way that looks random but actually brings the cards back to their original order. Do not gamble with this man.)

If a coin flip is deterministic physics masquerading as randomness, maybe the decay of a radioactive nucleus is just as predictable, under the surface—the only difference being that we don't yet know the secret. In quantum mechanics this kind of idea is known as a hidden-variable theory. It suggests a deeper level of activity, somewhere beyond the reach of present experimental measurement, controlling the behavior of the subatomic particles we observe. If we knew the values of the hidden variables, nothing in nature would seem random or causeless.

Some people find such secret mechanisms an appealing alternative to a world where stuff just happens willy-nilly. This is what Einstein had in mind when he proclaimed that God does not play dice. However, the vast majority of physicists disagree with Einstein on this point. Getting rid of randomness by invoking hidden variables comes at a high price. It seems you have to give up another kind of causality, namely the principle that information can travel only from the past to the future, and never faster than the speed of light. If hidden variables exist, they bring with them something else that Einstein campaigned to expunge from modern physics: "spooky action at a distance." I don't think we've quite heard the end of the hidden-variables debate, but for now the smart money is on real randomness.

CHAPTER 3

Follow the Money

The rich get richer and the poor get poorer. You've heard that before. It is a maxim so often repeated, and so often confirmed by experience, that it begins to sound like a law of nature, as familiar and irresistible as gravity. And indeed perhaps there is some physical or mathematical rule governing the distribution of wealth in the world. No such general principle is going to explain the specifics of *who* gets rich and poor, but maybe there's a way to understand the overall shape of the curve.

This idea goes back at least a century to the work of the Italian economist Vilfredo Pareto, who tried to show that the income distribution in all cultures and countries has the same mathematical form. In recent years the topic has been taken up with renewed enthusiasm by a small band of "econophysicists," who apply principles of statistical mechanics to questions in economic theory. The essence of their approach is to study an economy as if it were a many-body physical system such as a gas. Just as random collisions between molecules in a gas give rise to macroscopic properties such as temperature and pressure, random encounters between individuals in an economic system might determine large-scale phenomena such as the distribution of wealth.

Some of the computational models for exploring these ideas are remarkably easy to build and run. It takes just a few minutes' effort and a few lines of code. On the other hand, it's also remarkably easy to make subtle mistakes, as I'll have occasion to mention below. And the big challenge is not building the models but interpreting the results—deciding which kinds of random encounters might represent events in a real economy.

The Price Is Right

The economy in these models is a rather special one, based on pure, free-market trading. All that ever happens is buying and selling; no one produces or consumes anything. It's an economy with an eBay and a Wal-Mart and maybe an *Antiques Roadshow* but no General Motors, or even a family farm. Leaving out so much of the real economy is an obvious weakness of the model, but there is a compensating advantage: what remains is a closed system. In the simulated economy, wealth is a conserved quantity, like energy or momentum in a physical system. Because the total amount of wealth never changes, one person can get richer only if another grows poorer. This simplifies the accounting.

I find it helpful to think of this miniature economy in terms of a yard sale, where all the participants put their goods out on

Wealth flows through a model economy in which vertical channels represent individual traders and horizontal channels show the gain or loss in transactions between traders. In this specific model (the "yard sale" economy), gain or loss is limited to the wealth of the poorer trading partner. All traders start out equal at the top of the diagram, but wealth becomes more concentrated with time.

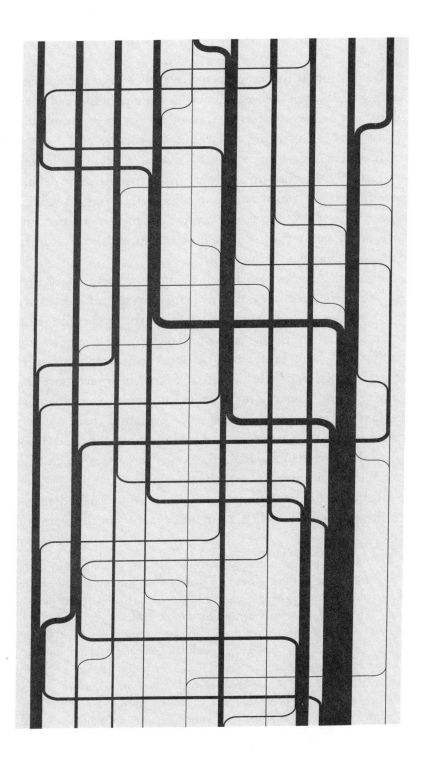

the lawn Saturday morning, then stroll up and down the street making trades with their neighbors. At the end of the day, after all transactions are completed, an auditor reviews everyone's inventory and calculates any changes in net worth.

Many economic theories assume that all transactions occur at precisely the right price. Indeed, prices are correct by definition in such theories: whatever price is agreed to by a willing seller and a willing buyer is the value assigned to an asset. In such an economy there can be no bargains and no bad deals, because the price is always right.

Given such perfect pricing, nothing interesting could ever happen in the yard-sale economy. I might trade my old toaster for your broken VCR, but if we negotiate the terms of the deal correctly, my net worth will not change in the slightest, and neither will yours. In practice, though, the assumption of perfect pricing seems a little unrealistic. Some buyers are more discerning or more patient than others, and some sellers are more persuasive. There are bargains to be had, and there are certainly bad deals—concepts that could not exist if we did not agree that merchandise has a true, intrinsic value, which does not necessarily correspond to the price paid.

Slight departures from perfect pricing bring a new dynamic to the yard-sale model. If I buy your rusty wheelbarrow and

---

The distribution of wealth in the yard-sale economy grows steadily more extreme with the passage of time. In each of these graphs the dashed line indicates the starting configuration: 1,000 traders, each possessing one unit of wealth. Curves in progressively darker shades of gray show the distribution after increasing numbers of transactions, from 1,000 up to 10 million. The upper graph plots wealth density: the total amount of money held by people at each wealth level. The lower graph is a Lorenz curve (named for the Italian economist Max O. Lorenz), which gives the proportion of the population that controls a given proportion of the wealth. In the end a single trader accumulates essentially all the wealth.

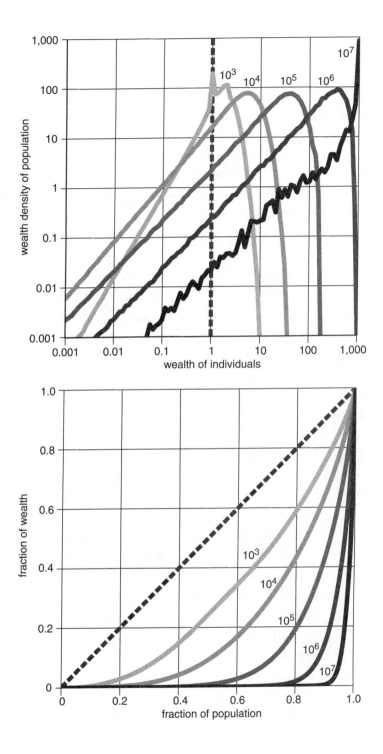

pay more than it's worth, I am left a little poorer after the transaction, and you are a little richer. Conversely, if I pay less than fair value, I gain a little, and you lose. In either case there has been a transfer of wealth, typically a small fraction of the price paid. These transfers are where all the action is in the modeled economy. As a matter of fact, the model can ignore the transaction itself—there's no need to talk about toasters and wheelbarrows—and simply consider the net transfer of wealth, which is equal to the discrepancy between the price paid and the true value.

The question is: What happens when this process is repeated many times? If some of the traders are shrewder than others, you would certainly expect them to do well in the long run; likewise, the perennial suckers are going to lose their shirts. But suppose that everyone is equally skillful, so that who wins and who loses in any particular transaction is purely a matter of chance. The amount of gain and loss is also determined at random—but it's always less than the total wealth of the poorer agent, so that traders never risk losing more than they own. What happens in such a toy economy?

Before reading on, you might try to predict the outcome. Assume that everyone starts out with the same bankroll. How will the assets be distributed after many random exchanges? Will the levels of wealth remain uniform? Perhaps the system will evolve toward a Gaussian, or normal, distribution, with most people having a middling amount of money while a few are very poor and a few are rich.

A computer program to simulate this process is very simple. An array of numbers records the wealth of each individual in the population; at the outset, all these numbers are equal. Then two individuals—two elements of the array—are chosen at random to participate in a transaction. The amount to be gained or lost is set to a random value no greater than the current wealth of the poorer participant. Then yet another random choice determines which party wins and which loses. After the debits and credits are posted to the array, the whole procedure starts over with the random choice of another pair of traders.

Running this program yields a stark result: if trading continues long enough, essentially all the wealth winds up in the hands of one person. The yard-sale economy, as formulated in this model, is a winner-take-all lottery. The traders might just as well put all their goods in one big pile, and then roll the dice to decide who keeps it all. (Strictly speaking, one trader gets *all* the goods only if wealth is quantized—only if there is some smallest unit of value below which one's worth falls to zero. If wealth can be subdivided indefinitely, the winner's share comes arbitrarily close to 100 percent but never quite gets there.)

This condensation of all property in the hands of one individual is an economic catastrophe—something like the formation of a black hole in astrophysics. It's obviously bad news for the majority of the people, who are left penniless. But even if you happen to be the big winner, your victory may prove hollow. Although you have all the riches in the world, you can't buy a thing, because no one else has goods to sell. And you can't sell anything either, because no one has money to buy with. The whole economy is frozen.

Molecular Economics

I first began experimenting with the yard-sale simulation after reading an article, "Wealth Distributions in Asset Exchange Models," by Slava Ispolatov, Paul L. Krapivsky, and Sidney Redner of Boston University. The computer models described there seemed both intriguing and easy to re-create, and so I wrote a quick-and-dirty program to play with some of them. I was perplexed when my results were quite different from those reported in the article. A second look revealed that I had misread a key equation, so that my model differed from theirs in a small but crucial way. Later I found a paper by Anirban Chakraborti of the Saha Institute of Nuclear Physics in India that describes essentially the same model I had accidentally created.

At least two other groups of physicists have published work on related themes. In France, Jean-Philippe Bouchaud of the Centre d'études de Saclay and Marc Mézard of the École normale supérieure have described "wealth condensation" in a somewhat different model. And Adrian Drăgulescu and Victor M. Yakovenko of the University of Maryland have written on the "statistical mechanics of money."

A source of ideas for most of these models is the analogy between market economics and the kinetic theory of gases. The molecules of a gas are continually colliding with one another and exchanging energy, in much the way that randomly chosen buyers and sellers in an economic model exchange sums of money. Yet gases do not follow the evolutionary path of the yard-sale economy. An economic collapse, where one person sucks in all the money, corresponds to a gas where one molecule has all the kinetic energy: it zips around at enormous speed while all the rest of the molecules are standing still. Don't hold your breath waiting for that to happen.

A subtle change in the economic model yields a totally different distribution of wealth. The change alters the rule for setting the amount of each transaction: the upper limit is the loser's wealth rather than the wealth of the poorer partner. Under these conditions no individual acquires more than a few percent of the total wealth. Moreover, the distribution is stable: after the first few thousand transactions, the shape of the curves does not change. Ironically, the kinds of transactions that produce this pleasant outcome could be interpreted as theft and fraud.

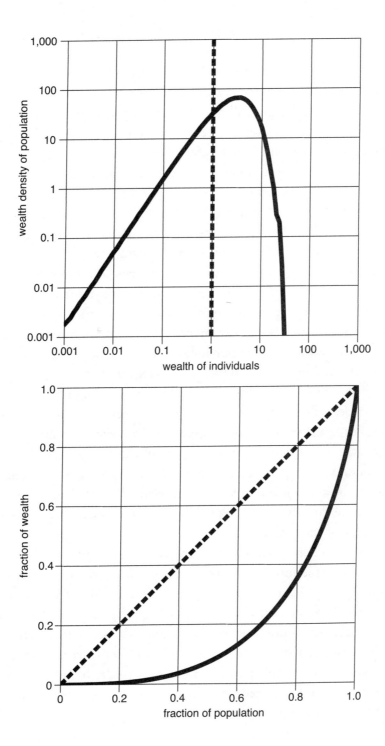

Where the yard-sale model departs from the kinetic theory of gases is in the details of the exchange of wealth or energy. When two molecules collide, they can reapportion their energy in any way that leaves the total unchanged. If the molecules have energies a and b just before they collide, then afterward they can have any combination of energies that add up to $a + b$. Translating this energy-redistribution process into financial terms yields a market in which the parties to every transaction lump together their entire fortunes and then randomly divide the total wealth before they go their separate ways.

A simulated economy based on this rule does not collapse the way the yard-sale model does; wealth remains spread throughout the population, although not uniformly so. The distribution follows an exponential curve: the number of people with wealth w is proportional to $1/e^{(w/T)}$, where e is the famous number 2.718 (known as Euler's number) and T is the temperature. Thus a larger w implies a smaller number of people possessing wealth w: the rich are fewer than the poor. A higher temperature implies a broader or flatter distribution of wealth. What does it mean to speak of the "temperature" of an economy? Drăgulescu and Yakovenko identify the temperature with the average amount of money available to the participants, much as the temperature of a gas measures the average kinetic energy of the molecules.

Although an exponential distribution of wealth crowds most of the people into the lower economic strata, the degree of inequity is mild compared with the all-or-nothing outcome of the yard-sale model. Furthermore, although the shape of the exponential distribution is stable, individuals do not remain stationary within it: there are many rags-to-riches-to-rags stories in such a society. And perhaps a gap between rich and poor will seem less unfair if people have a reasonable chance of moving between the two categories.

An exponential distribution of wealth is clearly preferable to a winner-take-all outcome, and an economic model based on the kinetic theory of gases may also have a certain aesthetic appeal—at least to physicists. Nevertheless, the interpretation of the model is problematic. There is no obvious reason to expect economic agents to act like colliding molecules, and indeed the random repartitioning of kinetic energy is a fairly strange template for mercantile transactions. Applied in the yard-sale context, it suggests that when Bill Gates comes to browse among my lawn ornaments, he and I will pool all our assets and then randomly split up the pot. Sounds good to me, but will Bill go along with that plan?

One kind of financial transaction that might fit the pattern of the kinetic theory of gases is marriage followed by divorce: this is a case where the parties *do* combine their holdings and later redivide them, although perhaps not quite randomly. In the corporate world, mergers and spin-offs might produce similar results.

Crime Doesn't Pay

The two models described so far lie at opposite poles along an axis defined by the amounts that the trading parties put at risk. In the yard-sale model, the most that can be won or lost is the total wealth of the poorer partner. Since this model evolves toward a state where nearly everyone is impoverished, the typical transaction is extremely small. In the marriage-and-divorce model, in contrast, the entire fortunes of both partners are up for grabs.

Here is a recipe for a third model that occupies a middle ground. As in the yard-sale algorithm, pick two trading partners at random, and also randomly choose which of the partners is to lose (the donor) and which is to gain (the recipient).

But instead of setting the size of the trade as a random fraction of the poorer player's wealth, make it a random fraction of the wealth of the donor. This rule still satisfies the commonsense constraint that you can never be made to pay more than you have. In each transaction you risk losing a random fraction of

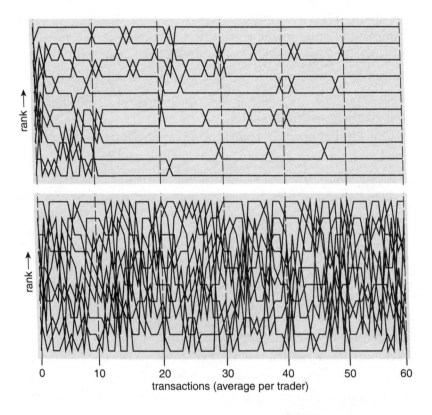

Economic mobility varies between models. In these diagrams time proceeds from left to right, and economic ranking increases from bottom to top. Each line represents the trajectory of a randomly selected trader. The yard-sale model (upper panel) is highly stratified: whoever gets to the top after the first few transactions has a good chance of staying there for a long time. In the theft-and-fraud model (lower panel) it's rags to riches and back again for everyone.

your own wealth, but you have a chance to gain a random fraction of the other person's fortune.

What kinds of real-world transactions might be described by this model? No doubt there are many plausible interpretations, but here is one that I find intriguing. A distinctive characteristic of the trading scheme is that the richer party always has more to lose and the poorer has more to gain. Under these terms, any sensible person would try to do business only with wealthier partners, and no one would ever willingly choose to trade with a less-affluent person (assuming traders can gauge the wealth of their partners). Thus if trading between non-equals takes place at all, it must be by coercion or deception. In other words, what is being modeled here is theft and fraud.

When the theft-and-fraud model is allowed to run for many iterations, there is no economic collapse. The wealth distribution reaches an equilibrium on an exponential curve much like that seen in the marriage-and-divorce model. (I have no comment on this evidence that marriage and divorce have the same economic impact as larceny, nor will I speculate on why a world populated by bank robbers winds up with a fairer distribution of wealth than an economy of honest merchants.)

Beyond the Dreams of Avarice

Recent publications on asset-exchange models describe many more variations. Drăgulescu and Yakovenko mention a family of models that differ among themselves only in the rule for choosing an amount of money to transfer. In one case it is a small fixed quantity; in another it is a random fraction of the trading pair's average wealth; in a third model the amount is a random fraction of the average wealth of the entire population. To avoid putting traders into debt or bankruptcy, Drăgulescu and Yakovenko apply the meta-rule that if the loser cannot pay, the transaction is canceled. In all these models the equilibrium distribution has an exponential form, and there is no collapse.

Ispolatov, Krapivsky, and Redner look at greedy or exploitative rules, where the wealthier party always wins the exchange (perhaps reflecting a situation where the poor have less bargaining power). When the amount transferred is a random fraction of the poorer agent's wealth (as in the yard-sale model), the result is economic collapse, with all funds gravitating toward one person. Of course it's no surprise that systematic greed yields a harsh outcome. The surprise is that this obviously biased rule is no worse than the symmetrical rule in the yard-sale model.

Chakraborti looks at the effect of savings, allowing traders to hold back some of their capital from the market. In the yard-sale economy, savings cannot forestall a collapse. Reserving a fixed sum of money shifts the minimum wealth up from zero but does not alter the dynamics of the model. Saving a fixed *fraction* of wealth slows the collapse, but the winner still takes all in the end.

Several authors mention the effects of taxes, welfare, and other explicit means of redistributing income. Imposing a tax on wealth prevents the implosion of the yard-sale economy, but the effects of an income tax are not so clear. I experimented with income taxes by collecting a percentage of each transaction and redistributing the proceeds in equal shares to all traders. A low tax rate does not protect against collapse, but models with tax rates higher than about 15 percent do seem to

---

Taxes and welfare forestall the collapse of the yard-sale model. Here the underlying transaction mechanism is the same as in the other simulations, but after every trade a randomly chosen trader is taxed a randomly chosen amount, and the proceeds are distributed in equal shares to all the traders. The result is a stable distribution with a comparatively narrow span between richest and poorest.

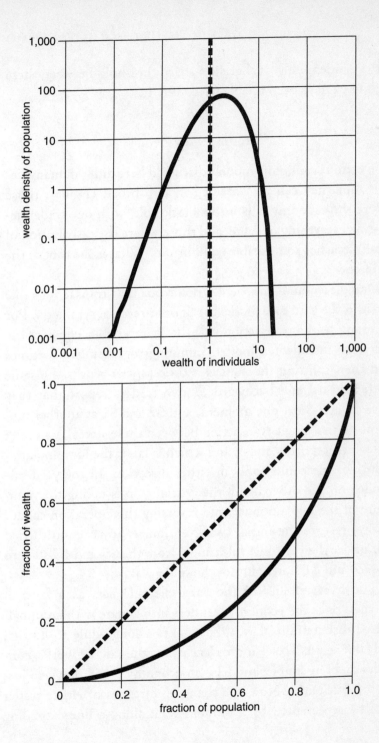

survive indefinitely. If there is a sharp threshold between these regimes, I have been unable to identify it.

Trading with Zeno

The various economic models discussed here differ in many details, but they can be classified in two broad families: those where the economy falls into a black hole, with one trader acquiring everything of value, and those where the distribution of wealth reaches some stable equilibrium. What is the root of the difference?

Drăgulescu and Yakovenko point out that transactions like those in the yard-sale model break time-reversal symmetry. For an example of a transaction rule that is reversible, consider the marriage-and-divorce model. Lumping together two fortunes and then splitting the sum is a process that works the same both forward and backward. If two traders report that they have $5 and $3 at one moment, and $7 and $1 at another moment (with a single transaction between these states), you can't tell which report is earlier and which is later; the lumping-and-splitting rule could apply in either direction. In the yard-sale model, on the other hand, the crucial step is taking the minimum of the two amounts, and reversing this operation cannot always restore the initial configuration. A transaction carried out under the yard-sale rule can go from the $5-and-$3 state to the $7-and-$1 state, but not the other way.

The irreversibility of the yard-sale rule acts as a kind of ratchet: once the economy wanders into a state with an unbalanced distribution of wealth, it takes a long while to find its way out again. To see more clearly how the ratchet works, consider an even simpler model—an economy pared down to just two participants. Now the changing fortunes of either trader can be represented by a random walk along a line extending from zero to the total wealth available. All activity stops if the trader reaches either end of this line. A random walk that takes steps of uniform length is guaranteed to hit an endpoint sooner or later (a fate known as gambler's ruin). But this is not what is going on in the yard-sale model. There the steps are not of fixed size; because transactions are limited to the lesser of the trading partners' assets, the steps get smaller as the walk approaches either endpoint. If there is no smallest unit of currency, the random walk becomes a "Zeno walk," named for the Greek philosopher who spoke of arrows that cover half the distance to the target, then half the remaining distance, then half again—never actually reaching the destination.

To simplify the model still further, we can take a Zeno walk on the interval from 0 to 1, choosing to go left or right at random but letting the step size always be half the distance to the nearer endpoint (rather than a random fraction of this distance). If we begin at the point ½, the initial step size is ¼. Suppose the first move is to the right, reaching the point ¾. Now the step size is ⅛. If we turn back to the left, we do not return to our starting point but instead stop at ⅝. Where will we wind up after n steps? The probability distribution for this process has an intricate fractal structure, so there is no simple answer, but the likeliest landing places get steadily closer to the endpoints of the interval as n increases. This skewed probability distribution is the ratchet-like mechanism that drives the yard-sale model to states of extreme imbalance.

Fair Trade

Models of the market economy may lead to some cute mathematics, but do they have the slightest connection with the price of peas in the real world? Can they predict the actual distribution of wealth observed in human societies?

As it happens, the shape of the actual distribution is uncertain and controversial. Most of the available data concern the distribution of *income*, which is not quite the same as the distribution of wealth. Pareto, a hundred years ago, argued that the income distribution obeys a power law, so that the proportion of people whose income is at least x varies as $1/x^a$; Pareto believed that the exponent a is a universal constant with a value of about 2.5. Other economists have proposed a "lognormal" income curve, meaning that the distribution of the logarithm of income follows a normal curve. A power law sets no definite upper limit on wealth, whereas a lognormal or exponential distribution has a sharp cutoff.

The model of Bouchaud and Mézard (which includes investment earnings as well as trade) yields a Pareto-like power law for the wealth distribution. Some of the "greedy" models of Ispolatov, Krapivsky, and Redner also appear to fit a power-law curve. But the models drawn most directly from the kinetic theory of gases predict an exponential distribution of wealth. Drăgulescu and Yakovenko argue that the middle part of the actual wealth distribution is indeed exponential, with a "Pareto tail" in the highest wealth brackets. All the computational models are so crude, however, and the empirical measurements are so uncertain, that curve fitting inspires little confidence.

Also unclear is whether events comparable to the collapse of the yard-sale model can happen in a real economy. Societies where a small elite controls almost all the property, while the rest of the people are destitute, are all too common. But does this situation result from a mathematical instability in the system of trade, or is there a simpler explanation, such as mere malice and greed? In any case, economic collapse seems never to go to completion in the real world. Tycoons amass immense fortunes, but no one ever goes home with *all* the marbles. (Bill Gates holds much less than 1 percent of the world's wealth.)

Rather than trying to match the output of the models to economic statistics, we might find it more fruitful to examine real-world economic practices for signs of the basic mechanisms that underlie the models. In particular, the fatal feature of the yard-sale model is the rule limiting the size of a transaction to the wealth of the lesser trading partner. The rule appears to be perfectly fair and symmetrical, and yet it has the effect that the further you fall through the economic strata, the harder you'll find it to climb back up.

Is such a rule likely to be enforced in everyday commerce? Not always. It is clearly violated in many forms of gambling and speculation, where the whole point of the transaction is the hope of gaining more than you put at risk. Doubtless there are other exceptions as well. For the most part, though, those of us with less money are limited to smaller-scale buying and selling. And the lower the ceiling on your economic activity, the slower your progress up through the ranks. When I buy a new car, I have little chance—no matter how shrewdly I bargain—of significantly altering the balance of assets between me and General Motors.

Explaining the distribution of wealth among individuals is not the only possible application of the trading models. They might in fact be better suited to describing relations among companies, where a sudden condensation of wealth could be interpreted as the emergence of a monopoly.

Beyond the corporate world, there is the question of whether the models might have anything to say about commerce among nations, and the ongoing debate over free markets, fair trade, and a "level playing field." If some mechanism like that of the yard-sale model is truly at work, then markets might very well be free and fair, and the playing field perfectly level, and yet the outcome would almost surely be that the rich get richer and the poor get poorer. You've heard that before.

AFTERTHOUGHTS

When this essay was published in *American Scientist* in 2002, several readers took issue with the very first line: "The rich get richer and the poor get poorer." No one disputed that the rich have been getting richer, but the second half of the conjunction met with skepticism in some quarters. One correspondent, Robert Lyman of Seattle, objected:

> Actually, the claim that poor people are made poorer by the modern market economy is blatantly false, and is repeated only by economically illiterate politicians bent on class warfare. The poor of 2002 are immensely better off than the poor of 1902, 1952, or even 1982 . . . The $10 all-you-can-eat buffet contains a variety and quantity of foods which would have amazed the robber barons of the early twentieth century—and they would have been even more amazed to see coal miners and factory workers able to afford such a feast several nights a week.

The assertion that people at the bottom of the economic ladder have climbed a rung or two deserves to be taken seriously. It is probably true. According to Shaohua Chen and Martin Ravallion of the World Bank, the mean income of people in the very lowest economic stratum rose from seventy cents a day in 1981 to seventy-seven cents in 2001 (in dollars of constant value). About a billion people, a sixth of the world's population, live on such a meager budget. Looking back over a much longer span of time, François Bourguignon and Christian Morrisson (also affiliated with the World Bank) find that the mean income of the poorest 20 percent of the world's population tripled between 1820 and 1992. Thus my glib truism about the poor getting poorer finds no support in these statistics. Although I would not characterize seventy-seven cents a day as "immensely better off," I concede that the trend appears to be upward. After a few more centuries, perhaps the poorest billion will even be able to afford the $10 buffet.

What I should have said, instead of relying on that pithy formula, is that the disparity between rich and poor has been growing wider. Over the two-century interval in which the poorest fifth of the population tripled their income, the wealthiest fifth gained by a factor of ten. For the mathematical models described in the article, it is only such differences or ratios of wealth that enter into the calculation. The models concern the *distribution* of wealth—how evenly or unevenly it is spread through a population. Shifts or rescalings that affect everyone's wealth in the same way are irrelevant to the models.

A closely related criticism of the essay focused on the "zero-sum" nature of the models. Thomas E. Moore of the NASA Goddard Space Flight Center wrote:

> I think the biggest problem is with the model assumption that wealth is a fixed or conserved quantity. The engine that drives our world economy is the creation of wealth by the collective action of people who work. By converting relatively low value materials into goods or services that others want, wealth is generated in huge quantities each and every day.

At one level, this complaint reflects a misunderstanding. It was never my contention that wealth really is a conserved quantity; there's no question that digging coal out of the ground or growing carrots creates something of value and thereby injects new wealth into the economic system. The models don't deny the existence of such productive activity; they just ignore it, in order to focus attention elsewhere. An analogy in physics would be a model of planetary motions that ignores the overall velocity of the solar system through the galaxy. No one doubts that such a velocity exists, but it's neither necessary nor helpful to include it in a simulation of orbital mechanics. Instead, we view the solar system as if it were standing still in space, and likewise we track the distribution of wealth through a society as if the total wealth were a constant.

A subtler view of this issue asks whether a model can so easily be separated into independent parts. In the case of the astronomical model, the planetary and galactic motions decouple perfectly in Newtonian physics, but they are only approximately independent in Einstein's relativity. In the economic context, I have assumed that transfers of wealth caused by small price discrepancies during trade will operate in the same way no matter where the wealth comes from in the first place. I see no reason to doubt this assumption, at least as an approximation, but I also have no evidence to support it.

Taking a step back from these detailed critiques, one can ask whether the whole notion of basing an economic model on the physics of gases ought to be taken seriously. Is this just mathematics at play, or can we learn something about real societies? The models are so simple that they cannot be expected to make reliable quantitative predictions. Still, there is abundant evidence for the existence of *some* mechanism that tends to concentrate wealth in the hands of a few. Many theories offer to explain that tendency. No doubt some people are rich because they worked harder to earn their wealth. Another factor is a well-known positive feedback loop: people with surplus capital to invest have opportunities for gain that are unavailable to the seventy-seven-cents-a-day crowd. The asset-exchange models suggest there may well be another feedback mechanism that tends to favor wealthier trading partners.

Since this essay appeared in *American Scientist*, there have been many further publications on the same theme. I have also learned of interesting earlier work that seems to have escaped the notice of the econophysics community. In 2005 Thomas Lux of the University of Kiel called attention to several papers published in the 1980s and '90s by John Angle, who was then a statistician with the U.S. Department of Agriculture. Angle studied the distribution of wealth by means of an interacting-particle system, a model closely related to those discussed in this essay, although Angle's system differs in that the richer party is explicitly given a higher probability of coming out ahead in each transaction.

A full century earlier, John Ruskin, the English critic and essayist, prefigured some of the ideas that emerge from the econophysics model. In *Unto This Last* he wrote:

Men nearly always speak and write as if riches were absolute, as if it were possible, by following certain scientific precepts, for everybody to be rich. Whereas riches are a power like that of electricity, acting only through inequalities or negations of itself. The force of the guinea you have in your pocket depends wholly on the default of a guinea in your neighbor's pocket. If he did not want it, it would be of no use to you; the degree of power it possesses depends accurately upon the need or desire he has for it,—and the art of making yourself rich, in the ordinary mercantile economist's sense, is therefore equally and necessarily the art of keeping your neighbor poor.

CHAPTER 4

Inventing the Genetic Code

As James Watson tells the story, on the last day of February in 1953, Francis Crick stood up and announced to the patrons of the Eagle pub in Cambridge: "We have discovered the secret of life." History supports the boast. If life ever had a secret, the double helix of DNA was surely it. And yet Watson and Crick had not laid bare all the secrets of molecular biology. The campaign to understand the code embodied in the double helix was just beginning, and the years ahead would be notable for frustration, false starts, and brilliant ideas that turned out to be utterly wrong. It took another full decade to solve the riddle.

Some time ago I found myself browsing in the literature of that curious decade. I had come upon one paper by chance while looking for something else, and I was so intrigued that I tracked down some of the earlier works it cited. A few days later I came back to peel away another layer of references. Then I shifted forward in time to read later summations and histories. (This kind of truffle hunting in the library stacks is especially engaging when you're supposed to be doing something else.)

What fascinated me about the code-breaking effort was how quickly a biochemical puzzle—the relation between DNA structure and protein structure—was reduced to an abstract problem in symbol manipulation. Within a few months, all the messy molecular complexities were swept away, and the goal was understood to be a mathematical mapping between messages in two different alphabets. The language of DNA has four letters: A, T, G, C, standing for the nucleotide bases adenine, thymine, guanine, and cytosine. Proteins are long chains of amino acids, which come in about twenty varieties. When biologists started speaking of a "genetic code," what they meant was a key for translating sentences written in the four-letter DNA alphabet into twenty-letter protein texts. Most of the tools for finding that key came not from biology or chemistry but from areas of mathematics and computer science. Proposed solutions were judged by how efficiently a code could store and transmit information—the kinds of criteria that an engineer might apply in designing a communications protocol for a computer network.

The early work on the genetic code fascinated me for another reason as well: some of the proposed codes were truly ingenious. Indeed, it was hard not to feel a twinge of regret on coming to the end of the story and learning the right answer. Compared with the elegant inventions of the theorists, nature's code seemed a bit of a kludge.

What We Didn't Know Then

To enter the world of molecular biology circa 1953, you must first forget all you know. This isn't easy when you come from a world where the sequencing of entire genomes is routine. In 1953 no one had yet read the sequence of bases in any DNA molecule—not one scrap of one gene.

For proteins the situation was only a little better. Frederick Sanger was finishing his work on the amino acid sequence of insulin, and a few other fragmentary protein sequences had been published. But the very idea that every protein has a precisely defined sequence, the same in all copies of the molecule, was not yet universally accepted. Even the set of amino acids from which proteins are assembled was still subject to dispute (although Watson and Crick would soon sit down at the Eagle to write out the canonical list of twenty). And all the biochemical apparatus for translating DNA into protein awaited discovery. The intermediate molecules called messenger RNA and transfer RNA were unknown. Ribosomes, the sites of protein assembly, had been glimpsed in electron micrographs, but their function was unclear.

One area that was not quite so murky was the replication of DNA. From the moment Watson and Crick saw that the four nucleotide bases fit together in specific pairs—adenine with thymine, guanine with cytosine—the mechanism of replication seemed obvious: unzip the double helix and form two new strands complementary to the original ones. One reason this process was so much easier to fathom was that the replication machinery does not have to consider the meaning of a base sequence in order to duplicate it, any more than a Xerox machine has to understand the documents it copies.

Translation, in contrast, cannot avoid semantics—and yet no one had a clue about how to interpret a sequence of bases in DNA. Even the most fundamental questions remained open. For example, since DNA is a *double* helix, should you look for information on both strands? If only one strand carries the message, how do you know which one it is? And which direction do you read in? Trying to make sense of the genome was like being given a book in a language so unfamiliar you couldn't be sure you were holding it right side up.

The Diamond Code

The first coding scheme inspired by the Watson-Crick structure came from an unexpected quarter. The author was not a biologist or a chemist but a physicist: George Gamow, the chief proponent of the big bang theory in cosmology.

In Gamow's initial proposal, which he called the diamond code, double-stranded DNA acted directly as a template for assembling amino acids into proteins. As Gamow saw it, the various combinations of bases along one of the grooves in the double helix could form distinctively shaped cavities into which specific amino acids might fit. Each cavity would attract a certain amino acid, depending on the identity of the surrounding nucleotide bases; when all the amino acids were lined up in the correct order along the groove, an enzyme would come along to polymerize them, linking them into a continuous chain.

Each of Gamow's cavities was bounded by the bases at the four corners of a diamond. If the DNA helix is oriented vertically, the bases at the top and bottom corners of a diamond are on the same strand and are separated by a single intervening base; the left and right corners of the diamond are defined by that intervening base and by its complementary partner on the opposite strand.

Some years later, Crick wrote: "The importance of Gamow's work was that it was really an abstract theory of coding, and was not cluttered up by a lot of unnecessary chemical details." Actually, Gamow's description of the diamond code had more chemical clutter than many of the later proposals, but it was indeed the abstract parts of the scheme that made an impression and had a lasting influence. In particular, Gamow's treatment of the problem of mismatched alphabets is still the starting point for textbook accounts of the genetic code.

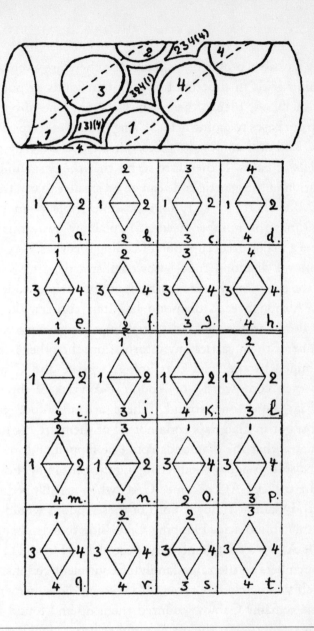

George Gamow's diamond code assumed that proteins form directly on a DNA template. In this 1954 drawing, nucleotide bases are designated by numbers, and the twenty codons specifying amino acids are given by letters. (Reprinted with permission from Nature, 173:318. Copyright Macmillan Magazines Ltd.)

The alphabet problem is simply that there are twenty kinds of amino acids in proteins but only four kinds of nucleotide bases in DNA. Hence there cannot be any one-to-one mapping from bases to amino acids. Using two bases to specify each amino acid still comes up short, since there are only sixteen doublets of bases. It therefore seems that the basic unit of information in the genetic code can be no smaller than a triplet of bases. But there are sixty-four possible triplets—more than three times the number needed. Explaining away this excess became a major preoccupation of coding theorists.

Gamow's diamond code—viewed abstractly, after sweeping away the chemical clutter—turns out to be a triplet code in disguise. Although each diamond has four corners, the paired bases along the horizontal diagonal are complementary, and so only one of them carries any information; the other is entirely determined by the rules that link A with T and C with G. Thus each code word—or "codon"—consists of three bases lined up along one strand. There are sixty-four possible codons, but in Gamow's code not all of them are distinct. He postulated that the diamonds could be flipped end for end or flopped side to side without changing their meaning. For example, the triplet CAG becomes GAC when it is flipped end for end, and Gamow thought both of these codons would specify the same amino acid. Flopping CAG side to side changes the middle A into a complementary T, so that CTG and GTC are also members of the same family of equivalent codons. When all such symmetries are taken into account, how many distinct codons remain? Gamow counted them up and found the answer is twenty—just the magic number he was looking for.

The diamond code had another important property: it was an *overlapping* triplet code. Each nucleotide base (except perhaps at the ends of a strand) claimed simultaneous membership in three adjacent codons. For example, the base sequence

GATTACA consists of five overlapping triplets: GAT, ATT, TTA, TAC, and ACA. At the time, overlapping triplets seemed like a good idea. There was a biochemical justification: the spacing between amino acids in a protein is similar to the spacing between bases in DNA, so that the two polymers mesh best when their subunits are matched one to one. The overlapping code also maximizes the density of information storage:

AAA	⟷	AUA		ACA	⟷	AGA
CAC	⟷	CUC		CCC	⟷	CGC
GAG	⟷	GUG		GCG	⟷	GGG
UAU	⟷	UUU		UCU	⟷	UGU

AAC	⟷	CAA	⟷	AUC	⟷	CUA
AAG	⟷	GAA	⟷	AUG	⟷	GUA
AAU	⟷	UAA	⟷	AUU	⟷	UUA
ACC	⟷	CCA	⟷	AGC	⟷	CGA
ACG	⟷	GCA	⟷	AGG	⟷	GGA
ACU	⟷	UCA	⟷	AGU	⟷	UGA
CAG	⟷	GAC	⟷	CUG	⟷	GUC
CAU	⟷	UAC	⟷	CUU	⟷	UUC
CCG	⟷	GCC	⟷	CGG	⟷	GGC
CCU	⟷	UCC	⟷	CGU	⟷	UGC
GAU	⟷	UAG	⟷	GUU	⟷	UUG
GCU	⟷	UCG	⟷	GGU	⟷	UGG

Symmetries of the diamond code sort the sixty-four codons into twenty classes. All the codons in each class specify the same amino acid. The sixteen palindromic codons at the top of the table come in pairs; the remaining codons are in groups of four.

even though three bases are needed to specify any single amino acid, the overall ratio of bases to amino acids approaches 1 to 1. Finally, overlapping imposes constraints on the possible sequences of amino acids. Gamow thought the constraints might reveal the nature of the true code; as it turned out, they were the downfall of his hypothesis.

The RNA Tie Club

A physicist popping up to tell biologists how to solve their problems can't always count on a warm reception. Gamow was welcomed, though, perhaps in part because biology labs in those days were full of carpetbagging physicists. (Crick himself began his career with a physics degree.) Or maybe Gamow just charmed his way in; by all accounts he was an exceptionally amiable fellow. In any case, he was soon spending a summer at the Marine Biological Laboratory on Cape Cod and collaborating with distinguished molecular biologists. He also founded the RNA Tie Club, limited to twenty regular members (one for each amino acid) and four honorary members (one for each nucleotide base). The ties were wool, with an embroidered green and yellow helix. Such an organization might not prosper today—who wears neckties?—but at the time it had an important role in circulating ideas.

The respect accorded to Gamow largely took the form of careful criticism. Attention focused particularly on his overlapping triplets. In any code where the ratio of bases to amino acids is 1 to 1, there are only 4^N nucleotide sequences of length N, but there are 20^N amino acid sequences. It follows that many of the amino acid sequences cannot be encoded by any base sequence. This effect can be seen even in an amino acid sequence of length 2 (called a dipeptide). With 20 kinds of amino acids, there are $20^2 = 400$ possible dipeptides, but 2 overlapping triplet codons comprise only 4 bases, so that there are only $4^4 = 256$ combinations. Evidently some 144 dipeptides cannot appear in proteins specified by an overlapping code.

Even with the sparse protein sequence data available in the mid-1950s, Crick was able to show that the diamond code was ruled out by the experimental evidence. There were known patterns of amino acid repetitions that the diamond code could not produce.

Undaunted, Gamow proposed a "triangle code" that was also overlapping but had different constraints. In this code, too, the sixty-four possible triplet codons sorted themselves into twenty families. Later, Gamow suggested yet another overlapping triplet code with an even simpler description: each codon is defined entirely by its base composition, ignoring the order of the bases within the codon. Thus ACT, ATC, CAT, CTA, TAC, and TCA are all members of the same codon family and specify the same amino acid. Remarkably, the number of codon families in this scheme again turns out to be exactly twenty. (It is just the number of combinations of four things taken three at a time.)

Still more overlapping codes came from Gamow and his friends. Richard Feynman had a hand in working out one idea. Edward Teller proposed another—a fairly funky scheme in which each amino acid is specified by two bases in the DNA and by the previous amino acid.

But overlapping codes were coming to the end of their string. Patterns of mutations were one source of doubt. With an overlapping code, changing a single base in the DNA could alter three neighboring amino acids, but protein sequence data were starting to show instances of single amino acid replacements. Then came a definitive proof. Sydney Brenner analyzed all the overlapping code

| A | G | A | C | G | A | U | U | A | U | C | A | A | C | A | G | C | C |

| A | G | A | C | G | A | U | U | A | U | C | A | A | C | A | G | C | C |

| A | G | A | C | G | A | U | U | A | U | C | A | A | C | A | G | C | C |

comma-free code

| A | G | A | C | G | A | U | U | A | U | C | A | A | C | A | G | C | C |

| A | G | A | C | G | A | U | U | A | U | C | A | A | C | A | G | C | C |

| A | G | A | C | G | A | U | U | A | U | C | A | A | C | A | G | C | C |

An overlapping genetic code packs sixteen codons (gray boxes) into a DNA strand of eighteen bases by assigning meaning to triplets of bases in all three "reading frames." Although it takes three bases to specify a single amino acid, each base is a member of three codons. A comma-free code is constructed so that only the codons in one reading frame are meaningful; the overlap triplets are nonsense (black boxes).

known protein sequence fragments and found enough correlations between nearest-neighbor amino acids to rule out every possible overlapping code.

In retrospect, the long fixation on overlapping codons seems unfortunate and misguided, but at the time there were strong arguments favoring such schemes. Matching the dimensions of the protein to those of the DNA template seemed important. So did coding efficiency. Natural selection was expected to maximize storage density and avoid any waste of information capacity. Engineers building the computers of the era certainly worked hard to pack in the bits, so why wouldn't nature do the same? No one could have guessed the awful truth—that nature is wildly profligate, that genomes are stuffed with gobs of "junk DNA," that storage efficiency just doesn't seem to be an issue except in a few ultracompact viruses.

Still another reason for favoring overlaps was to avoid the frame-shift problem. To understand the nature of this problem, it's best to turn to a very different kind of proposed code—one that I would like to nominate as the prettiest wrong idea in all of twentieth-century science.

Comma-Free Codes

By the later 1950s there was growing support for the idea of messenger RNA—a single-strand molecule acting as an intermediary between DNA and the protein-synthesizing machinery. At the same time Crick was formulating the "adaptor hypothesis," the idea that amino acids do not interact directly with messenger RNA but are carried by small molecules that recognize specific codons. (Today the adaptor molecules have been identified as transfer RNAs.) The codons were by then thought to be nonoverlapping triplets of bases.

The process of gene expression was imagined as going something like this: First the appropriate segment of DNA was transcribed into messenger RNA; like replication, this was done by blind copying, without regard to the meaning of the sequence. Then the messenger RNA stretched out in the cytoplasm of the cell with its long row of triplet codons exposed like a sow's nipples. Each adaptor molecule, already charged with the correct amino acid, poked around until it latched onto the right codon. When all the codons were occupied, the amino acids were linked together, and the completed protein was peeled off the template.

The scenario must have seemed highly plausible. Even from our current perspective, it seems like the kind of chemistry that living organisms do. The nonsequential pattern matching needed to line up adaptor molecules on the messenger RNA is vaguely like an enzyme-substrate reaction or like the binding of an antibody to an antigen. And yet there was a serious problem with the vision of piglets suckling on messenger RNA: a piglet might very well wind up between nipples, or straddling a set of nucleotide bases that belong to different codons.

Suppose somewhere in a strand of messenger RNA is the partial sequence UGUCGUAAG (note that in RNA uracil replaces the thymine of DNA, and so the code is written with U rather than T). The intended reading is UGU, CGU, AAG, but the RNA molecule has no punctuation to indicate codon boundaries. The sequence could equally well be read as UG, UCG, UAA, G or as U, GUC, GUA, AG. Each of these alternative "reading frames" would have a different meaning. Furthermore, in the suckling-pig model of protein synthesis, adaptor molecules that attached to the messenger RNA in different reading frames might interfere with one another and prevent any protein at all from being produced.

The frame-shift problem doesn't arise with an overlapping code, because all three reading frames are simultaneously valid. With sequential codons, however, the translation machinery has to be guided to the right frame. In 1957 Crick devised a solution that seemed at once so clever and so obvious that it just had to be right. He suggested that adaptor molecules might exist for only a subset of the sixty-four codons, with the result that only that subset would be meaningful; the rest of the triplets would be "nonsense codons." Then the trick is to construct a code in such a way that when any two meaningful codons are put next to each other, the frame-shifted overlap codons are always nonsense. For example, if CGU and AAG are sense codons, then GUA and UAA must be nonsense, because they appear inside the concatenated sequence CGUAAG. Similarly, AGC and GCG are ruled out by the sequence AAGCGU. If all the out-of-frame triplets are nonsense, then the message has only one reading, and adaptor molecules can attach to the messenger RNA in only one way. A code with this property is called a comma-free code, since it needs no punctuation, and messages remain unambiguous even when the words are run togetherwithoutcommasorspaces.

Do comma-free codes exist? In English you might try to find a subset of all three-letter words that can be jammed together without creating any additional instances of the words in the subset. To make the problem more manageable, consider this list of ten three-letter words: "ass," "ate," "eat," "sat," "sea," "see," "set," "tat," "tea," "tee." Is there a subset that forms a comma-free language? Trial and error shows that the words "ate," "eat," and "tea" cannot all appear together, because "teatea," for example, contains both "eat" and "ate." Similarly, "sea" combines with "tat," "tea," or "tee" to produce "eat." One set of words that has no conflicts is "ass," "sat," "see," "set," "tat," "tea," and "tee."

How many words can a comma-free code include? For the case of RNA, Crick and his Cambridge colleagues John Griffith (another physicist) and Leslie Orgel carried out a straightforward analysis. They pointed out first that the codons AAA, CCC, GGG, and UUU cannot appear in any comma-free code, since they cannot combine with themselves without generating reading-frame ambiguity. The remaining sixty codons can be sorted into groups of three, where the codons within each group are related by a cyclic permutation. For example, the codons AGU, GUA, and UAG form one such group. A comma-free code can have no more than one codon from each of these permutation classes. How many classes are there? Dividing sixty objects into groups of three produces exactly twenty groups. Bingo!

The analysis just given sets the maximum possible size of a comma-free genetic code, but it does not guarantee that a maximal code actually exists. Nevertheless, Crick, Griffith, and Orgel went on to construct several examples. And they offered a vision of how the code might work:

This scheme . . . allows the intermediates to accumulate at the correct positions on the template without ever blocking the process by settling, except momentarily, in the wrong place. It is this feature which gives it an advantage over schemes in which the intermediates are compelled to combine with the template one after the other in the correct order.

Crick and his colleagues were quick to point out that they had no experimental evidence for the comma-free code. As a nonoverlapping code, it put no constraints on amino acid sequences, so there was no point in looking for confirmation

AAA	CCC	GGG	UUU		

AAC	ACA	CAA		AUG	UGA	GAU
AAG	AGA	GAA		AUU	UUA	UAU
AAU	AUA	UAA		CCG	CGC	GCC
ACC	CCA	CAC		CCU	CUC	UCC
ACG	CGA	GAC		CGG	GGC	GCG
ACU	CUA	UAC		CGU	GUC	UCG
AGC	GCA	CAG		CUG	UGC	GCU
AGG	GGA	GAG		CUU	UUC	UCU
AGU	GUA	UAG		GGU	GUG	UGG
AUC	UCA	CAU		GUU	UUG	UGU

To build a comma-free code, first exclude the triplets AAA, CCC, GGG, and UUU, then divide the remaining sixty triplets into groups of three, where the triplets in each group are related by a cyclic permutation. A code can include no more than one triplet from each group. One comma-free code consists of the twenty triplets shown in gray boxes; all triplets in black boxes are nonsense.

there. The code did strongly constrain the base sequences of DNA and RNA, but those sequences were unknown. "The arguments and assumptions which we have had to employ to deduce this code are too precarious for us to feel much confidence in it on purely theoretical grounds," they wrote. "We put it forward because it gives the magic number—twenty—in a neat manner and from reasonable physical postulates." The magic number was enough to persuade both biologists and the wider public. Carl Woese later wrote:

> The comma-free codes received immediate and almost universal acceptance . . . They became the focus of the coding field, simply because of their intellectual elegance and the appeal of their numerology . . . For a period of five years most of the thinking in this area either derived from the comma-free codes or was judged on the basis of compatibility with them.

The intellectual elegance also attracted the attention of coding-theory professionals, most notably Solomon W. Golomb, now at the University of Southern California. Golomb and his colleagues (including the physicist-biologist Max Delbrück) wrote several papers on comma-free codes, taking the biological problem as their point of departure but going on to explore more abstract and generalized ideas. They quickly deduced a formula for the maximum size of a comma-free code: for an alphabet of n letters grouped into k-letter words, the formula takes a particularly simple form when k is a prime number: $(n^k - n)/k$. For $n = 4$ and $k = 3$ (the case of interest to biologists), the formula indicates a maximum code size of twenty, as expected. Golomb and his colleagues showed that there are 408 such maximal comma-free codes, and they gave a procedure for constructing them. They also devised some more-elaborate related codes. For example, a transposable comma-free code is designed so that both strands of the DNA have the comma-free property. Using triplets, the largest transposable code has only ten codons, but a quadruplet code yields twenty. Golomb also invented a genetic code based on sextuplets; not only is it comma-free and transposable, but also it can correct any two simultaneous errors in translation and detect the presence of a third error. Life would be a lot more reliable if Solomon Golomb were in charge.

Reality Intrudes

The comma-free codes were not quite the last word in the wildcat era of genetic-code building. In 1959 Robert Sinsheimer suggested a scheme in which the genetic alphabet had only two letters; A and C were interpreted as the same symbol, and so were G and U. This device was a way of coping with the recent discovery of wide variations in the ratio of $(A + U)$ to $(G + C)$ in various organisms. Of course reducing the code to binary notation meant that triplets of bases could not code for twenty amino acids; the codons would have to be at least quintuplets (providing thirty-two combinations).

As far as I know, no one ever proposed a three-letter (or ternary) code. Such a code might distinguish A from U but lump together G and C, producing twenty-seven codons. This plan has a faint echo in the real genetic code, where the third base in a codon is sometimes interpreted merely as A or G versus U or C.

I'm also surprised that no one gave serious thought to schemes where the codons can vary in length. In engineering, the idea of choosing shorter sequences to represent more frequent symbols was already a well-established trick for compressing a message. David Huffman had created a theory of such codes in 1951; the basic idea goes back at least a century further, to the Morse code for telegraphy. Biologists were clearly aware of the principle, and they were mindful of coding efficiency, but they did not explore the possibility.

Perhaps if the era of speculation had continued a few years more, these wrong ideas would also have been given their turn.

ACC	ACA	AAC	AAA	CAA	CAC	CCA	CCC
Thr	Thr	Asn	Lys	Gln	His	Pro	Pro
ACU	**ACG**	**AAU**	**AAG**	**CAG**	**CAU**	**CCG**	**CCU**
Thr	Thr	Asn	Lys	Gln	His	Pro	Pro
AUC	**AUA**	**AGC**	**AGA**	**CGA**	**CGC**	**CUA**	**CUC**
Ile	Ile	Ser	Arg	Arg	Arg	Leu	Leu
AUU	**AUG**	**AGU**	**AGG**	**CGG**	**CGU**	**CUG**	**CUU**
Ile	Met	Ser	Arg	Arg	Arg	Leu	Leu
GUU	**GUG**	**GGU**	**GGG**	**UGG**	**UGU**	**UUG**	**UUU**
Val	Val	Gly	Gly	Trp	Cys	Leu	Phe
GUC	**GUA**	**GGC**	**GGA**	**UGA**	**UGC**	**UUA**	**UUC**
Val	Val	Gly	Gly	⬡	Cys	Leu	Phe
GCU	**GCG**	**GAU**	**GAG**	**UAG**	**UAU**	**UCG**	**UCU**
Ala	Ala	Asp	Glu	⬡	Tyr	Ser	Ser
GCC	**GCA**	**GAC**	**GAA**	**UAA**	**UAC**	**UCA**	**UCC**
Ala	Ala	Asp	Glu	⬡	Tyr	Ser	Ser

Nature's own genetic code lacks the conspicuous symmetries that figured in most of the conjectured codes of the 1950s. The table gives the abbreviated name of the amino acid specified by each codon. Some amino acids have just a single codon; some have as many as six. Three of the codons (black boxes) are signals to stop protein synthesis. The evolutionary forces favoring this particular genetic code over all other possibilities are not obvious, but it's the code we live by all the same.

But in 1961 the whole coding craze was brought up short by unexpected news from the lab bench. Marshall W. Nirenberg and J. Heinrich Matthaei of the National Institutes of Health announced that artificial RNAs could stimulate protein synthesis in a cell-free system. What's more, the first RNA they tried was poly-U, a long chain of repeating uracil units. In comma-free codes, UUU has to be a nonsense codon, but Nirenberg and Matthaei's result implied that it codes for the amino acid phenylalanine. A few more codons were identified over the next year or two. Then Philip Leder and Nirenberg found an even better experimental protocol, and by 1965 the genetic code was mostly solved.

The code resembled none of the theoretical notions. As the table assigning codons to amino acids was filled in, it became apparent that the magic number twenty held no magic after all. All the clever mathematical contrivances for getting twenty amino acids out of sixty-four codons turned out to be figments of the human urge to find pattern, not reflections of any natural order. The "extra" codons are merely redundant: some amino acids have one or two codons, some have four, some have six. (Three codons serve as stop signs.) At first glance the mapping between codons and amino acids appeared arbitrary, even haphazard.

Nature also ignored all the mathematical ingenuity applied to solving the frame-shift problem. The living cell does it by a kind of dead reckoning. Ribosomes—the protein-making molecular machines—march along the messenger RNA in strides of three bases, translating as they go. Except for signals that mark where the ribosome is supposed to start, there is nothing in the code itself to enforce the correct reading frame.

When I mentioned to a biologist friend that I find some of the hypothetical genetic codes of the 1950s more appealing than the real thing, she protested that the actual code is one of the most elegant creations of biochemistry, and she pointed out some of its subtle refinements. The codon table is not entirely arbitrary. Its redundancies confer a kind of error tolerance, in that many mutations convert between synonymous codons. When a mutation *does* alter the meaning of the sequence, so that a protein winds up with an incorrect amino acid, the substitute is likely to have properties similar to those of the original, reducing the impact of the error. Computer simulations by David Haig and Laurence D. Hurst show that the present code is nearly optimal in this respect.

These observations suggest that I should be grateful my genes were not designed by George Gamow or Francis Crick. With Gamow's overlapping codes, any mutation could alter three adjacent amino acids at once, probably disabling the protein. Comma-free codes are even more brittle in this respect, since a mutated codon is likely to become nonsense and terminate translation. Most often, it's better to have a slightly defective protein molecule rather than none at all.

But these criticisms of the early conjectural codes are not entirely fair. They pluck the invented code out of its theoretical context and plug it into a biochemical system that has been evolving for three billion years or more in concert with a very different code. It's like replacing a man's arms with the wings of a bird and expecting him to fly. To truly test the merits of a comma-free code, we'd need to find a planet where the entire system of life was built on such a scheme. In that alien biochemical environment, we would doubtless find that our own code was maladaptive.

Imagine that in 1957 a clairvoyant biologist offered as a hypothesis the exact genetic code and mechanism of protein synthesis understood today. How would the proposal have been received? My guess is that *Nature* would have rejected the paper. "This notion of the ribosome ratcheting along the messenger RNA three bases at a time—it sounds like a computer reading a data tape. Biological systems don't work that way. In biochemistry we have templates, where all the reactants come together simultaneously, not assembly lines where machines are built step-by-step."

The Sixty-four-Codon Question

I want to conclude with a question. At the origin of life, the primitive genetic code was surely smaller and simpler than the modern one. It probably included only a few amino acids, or perhaps a few classes of similar amino acids. At some point in its history the code may have functioned as a pure doublet code, ignoring the third base in each codon and specifying no more than sixteen amino acids. Then the translation mechanism grew more discriminating, and a few more amino acids were added to the repertory. My question is: Why did this process of differentiation stop at twenty amino acids? There are plenty of spare codons left, and there are other amino acids that need to be gotten into proteins. So why not expand the code further?

One possible answer is that the code is such a vital engine of life that it has been immutable since the earliest stages of evolution. Tinkering with a mechanism so crucial to every cell's survival is too costly. Another answer is that the code *is* evolving steadily toward greater complexity, and we just happened to have discovered it at the twenty-amino-acid stage. Maybe our descendants will have sixty kinds of amino acids in their proteins. It's worth noting that twenty does not seem to be a hard-and-fast limit. The codon UGA, which is usually a stop signal, sometimes codes for a twenty-first amino acid, selenocysteine.

A third possibility is that there really is something special about the numbers sixty-four and twenty. The relation can't be the kind of numerological magic invoked by the comma-free codes, but perhaps there is some property of genetic codes that is optimized when the ratio of amino acids to codons approaches 1 to 3.

AFTERTHOUGHTS

The table with sixty-four squares, matching up codons with amino acids, has been printed in biology textbooks for forty years, and yet controversy over the meaning of the genetic code has never abated. Some of the issues discussed in this article have remained lively enough that I eventually wrote a sequel. ("Ode to the Code" appeared in *American Scientist* in 2004; you'll find it on the Web at www.americanscientist.org /AssetDetail/assetid/37228.)

Surprisingly, many of the more recent studies of the code continue in the same spirit as those of the 1950s. The early theorists of the code were sure that it would be the best of all possible codes in one way or another. The overlapping codes of Gamow and others had the highest possible information density. The comma-free codes were ideal for avoiding frame-shift errors in transcription. And almost all the proposals offered some compelling explanation of the numerical mystery—why sixty-four codons specify exactly twenty amino acids.

In hindsight, none of these factors seem very important; they are not the forces that have been driving the evolution of the genetic code. But biologists continue to seek signs that the natural code is optimal in *some* respect. The work by David Haig

and Laurence Hurst suggesting that the code is especially toler-
ant of mutations and mistranslations has been followed up
with many further experiments and computer simulations, notably by Stephen J. Freeland and his colleagues and students at
the University of Maryland, Baltimore County. They generated a million random codes—keeping the same sixty-four
codons and the same twenty amino acids, but reshuffling the
table that assigns each codon to an amino acid. They found
that among all these random permutations, the real biological
code is "one in a million." Mutations cause less harm when the
DNA is interpreted according to the real code than with any of
the random variations. Guy Sella of the Hebrew University of
Jerusalem and David H. Ardell of Uppsala University have
simulated the evolution of genetic codes by a different method
and reached similar conclusions.

More recently, Shalev Itzkovitz and Uri Alon of the Weiz-
mann Institute of Science have argued that the genetic code is
optimal in another way: it is ideal, they say, for carrying multiple messages at the same time. It has long been known that
DNA is not just an archive of recipes for making proteins.
There are lots of other meaningful sequences within the DNA,
such as binding sites for proteins that enhance or suppress the
expression of particular genes, and markers indicating where
RNA is to be cut or spliced. Thus a given stretch of DNA
might carry multiple meanings, as if it were a text that could be
read in both English and French. Obviously, the superimposed
messages could interfere with one another. Itzkovitz and Alon
find through computer simulations that the biological code is
one of the least susceptible to such interference.

The search for symmetries and patterns in the genetic code
also goes on. For example, Miguel A. Jiménez-Montaño, Carlos R. de la Mora-Basáñez, and Thorsten Pöschel have suggested organizing the codon table not as a $4 \times 4 \times 4$ cube in

three-dimensional space but as a $2 \times 2 \times 2 \times 2 \times 2 \times 2$ "hyper-cube" in six-dimensional space. Each of the sixty-four codons occupies a corner of the hypercube, and a mutation is a move-ment to one of the adjacent corners. Mark White, a physician in Indiana, has proposed a quite different geometry, mapping the twenty amino acids onto the faces of an icosahedron. (White has made valiant efforts to explain his ideas to me in more detail, but I still have only a tenuous understanding.)

The old dream of finding some deep connection between the numbers sixty-four and twenty has also never died. Pierre Béland and T.F.H. Allen have an idea that depends on a sup-posed ancient symmetry between the two complementary strands of the DNA double helix. If both strands are expected to carry meaningful information, then it turns out that twenty amino acids is the most that sixty-four bidirectional codons can supply. Although the modern cellular machinery generally reads only one strand of the DNA, Béland and Allen suggest that the earliest genetic code was double-stranded, and the lim-itation to twenty amino acids is a surviving artifact of that era.

All of these ideas are highly creative. A few of them are also supported by substantial evidence from experiments or simula-tions—in contrast to the more speculative flights of fancy from the 1950s. Still, the lessons learned from those early, uncon-strained attempts to decipher the code are surely worth keep-ing in mind. We are easily seduced by the beauty of our own ideas, but nature often ignores our aesthetic advice. In biology it's hard to resist seeing Darwinian evolution as a series of "just so" stories, in which natural selection always finds the best so-lution to every problem. In fact there is no such guarantee of optimality—at least not unless you turn Darwinism into a tau-tology and declare that whatever survives *must* be best. The ge-netic code we live by may indeed be unbeatable, but that needs to be established by evidence, not assumed as a given.

CHAPTER 5

Statistics of Deadly Quarrels

Look upon the phenomenon of war with dispassion and detachment, as if observing the follies of another species on a distant planet: from such an elevated view, war seems a puny enough pastime. Demographically, it hardly matters. Combatant deaths amount to something like 1 percent of all deaths; in many places, more die by suicide, and still more in accidents. If saving human lives is the great desideratum, then there is more to be gained by the prevention of drowning and auto wrecks than by the abolition of war.

But no one on this planet sees war from such a height of austere equanimity. Even the gods on Olympus could not keep from meddling in earthly conflicts. Something about the clash of arms has a special power to rouse the stronger emotions—pity and love as well as fear and hatred—and so our response to battlefield killing and dying is out of all proportion to its rank in tables of vital statistics. When war comes, it muscles aside the calmer aspects of life; no one is unmoved. Most of us choose one side or the other, but even among those who merely want to stop the fighting, feelings run high. ("Antiwar militant" is no oxymoron.)

The same inflamed passions that give war its urgent human interest also stand in the way of scholarly or scientific understanding. Reaching impartial judgment about rights and wrongs seems all but impossible. Stepping outside the bounds of one's own culture and ideology is also a challenge—not to mention the bounds of one's time and place. We tend to see all wars through the lens of the current conflict, and we mine history for lessons convenient to the present purpose.

One defense against such distortions is the statistical method of gathering data about many wars from many sources, in the hope that at least some of the biases will balance out and true patterns will emerge. It's a dumb, brute-force approach and not foolproof, but nothing else looks more promising. A pioneer of this quantitative study of war was Lewis Fry Richardson, a British meteorologist who set aside his work on weather prediction to study the mathematics of armed conflict.

Wars and Peaces

Richardson was born in 1881 to a prosperous Quaker family in the north of England. He studied physics with J. J. Thomson at Cambridge, where he developed expertise in the numerical solution of differential equations. Such approximate methods are a major mathematical industry today, but at that time they were not a popular subject or a shrewd career choice. After a series of short-term appointments—well off the tenure track— Richardson found a professional home in weather research, making notable contributions to the theory of atmospheric turbulence. Then, in 1916, he resigned his post to serve in France as a driver with the Friends' Ambulance Unit. Between tours of duty at the front, he did most of the calculations for a trial weather forecast by numerical methods. (The forecast was not a success, but the basic idea was sound, and all modern weather prediction relies on similar techniques.)

After the war, as Richardson gradually shifted his attention from meteorology to questions of war and international relations, he found some of the same mathematical tools still useful. In particular, he modeled arms races with differential equations. The death spiral of escalation—where one country's arsenal provokes another to increase its own armament, whereupon the first nation responds by adding still more weapons—has a ready representation in a pair of linked differential equations. Richardson showed that an arms race can be stabilized only if the "fatigue and expense" of preparing for war are greater than the perceived threats from enemies. This result is hardly profound or surprising, and yet Richardson's analysis nonetheless attracted much comment (mainly skeptical), because the equations offered the prospect of a quantitative measure of war risks. If Richardson's equations could be trusted, then observers would merely need to track expenditures on armaments to produce a war forecast, analogous to a weather forecast.

Mathematical models of arms races have been further refined since Richardson's era, and they had a place in policy deliberations during the "mutually assured destruction" phase of the Cold War. But Richardson's own investigations turned in a somewhat different direction. A focus on armaments presupposes that the accumulation of weaponry is a major cause of war, or at least has a strong correlation with it. Other theories of the origin of war would emphasize different factors—the economic status of nations, say, or differences of culture and language, or the effectiveness of diplomacy and mediation. There is no shortage of such theories; the problem is choosing among them. Richardson argued that theories of war could and should be evaluated on a scientific basis, by testing them against data on actual wars. So he set out to collect such data.

He was not the first to follow this path. Several lists of wars were drawn up in the early years of the twentieth century, and two more war catalogs were compiled in the 1930s and '40s by the Russian-born sociologist Pitirim A. Sorokin and by Quincy Wright of the University of Chicago. Richardson began his own collection in about 1940 and continued work on it until his death in 1953. His was not the largest data set, but it was the best suited to statistical analysis.

Richardson published some of his writings on war in journal articles and pamphlets, but his ideas became widely known only after two posthumous volumes appeared in 1960. The work on arms races is collected in *Arms and Insecurity*; the statistical studies are in *Statistics of Deadly Quarrels*. In addition, a two-volume *Collected Papers* was published in 1993. Most of what follows comes from *Statistics of Deadly Quarrels*. I have also leaned heavily on a 1980 study by David Wilkinson of the University of California, Los Angeles, which presents Richardson's data in a rationalized and more readable format.

Thinginess Fails

The catalog of conflicts in *Statistics of Deadly Quarrels* covers the period from about 1820 until 1950. Richardson's aim was to count all deaths during this interval caused by a deliberate act of another person. Thus he includes individual murders and other lesser episodes of violence in addition to warfare, but he excludes accidents and negligence and natural disasters. He also decided not to count deaths from famine and disease associated with war, on the grounds that multiple causes are too hard to disentangle. (Did World War I "cause" the influenza epidemic of 1918–19?)

The decision to lump together murder and war was meant to be provocative. To those who hold that "murder is an abominable selfish crime, but war is a heroic and patriotic adventure," Richardson replies: "One can find cases of homicide which one large group of people condemned as murder, while another large group condoned or praised them as legitimate war. Such things went on in Ireland in 1921 and are going on now in Palestine." (It's depressing that his examples, more than fifty years later, remain so apt.) But if Richardson dismissed moral distinctions between various kinds of killing, he acknowledged methodological difficulties. Wars are the province of historians, whereas murders belong to criminologists; statistics from the two groups are hard to reconcile. And the range of deadly quarrels lying between murder and war is even more problematic. Riots, raids, and insurrections have been too small and too frequent to attract the notice of historians, but they are too political for criminologists.

For larger wars, Richardson compiled his list by reading histories, starting with the *Encyclopaedia Britannica* and going on to more diverse and specialized sources. Murder data came from national crime reports. To fill in the gap between wars and murders, he tried interpolating and extrapolating and other means of estimating, but he acknowledged that his results in this area were weak and incomplete. He mixed together civil and international wars in a single list, arguing that the distinction is often unclear.

An interesting lesson of Richardson's exercise is just how difficult it can be to extract consistent and reliable quantitative information from the historical record. It seems easier to count inaccessible galaxies or invisible neutrinos than to count wars that swept through whole nations just a century ago. Of course some aspects of military history are always contentious; you can't expect all historians to agree on who started a war, or who won it. But it turns out that even more basic facts—Who were the combatants? When did the fighting begin and end? How many died?—can be remarkably hard to pin down. Lots of wars merge and split, or have no clear beginning or end. As Richardson remarks, "Thinginess fails."

In organizing his data, Richardson borrowed a crucial idea from astronomy: he classified wars and other quarrels according to their *magnitude*, the base-10 logarithm of the total number of deaths. Thus a terror campaign that kills 100 has a magnitude of 2 (since $10^2 = 100$), and a war with 1 million casualties is a magnitude-6 conflict ($10^6 = 1,000,000$). A murder with a single victim is magnitude 0 ($10^0 = 1$). The logarithmic scale was chosen in large part to cope with shortcomings of available data; although casualty totals are seldom known pre-

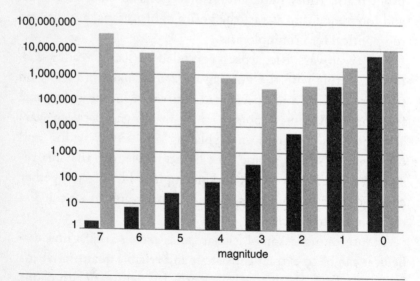

The magnitude of a war, as defined by Richardson, is the base-10 logarithm of the number of deaths. Black bars indicate the number of deadly quarrels between 1820 and 1950 in each magnitude range; gray bars are the total deaths from quarrels of that magnitude. Two magnitude-7 wars account for 60 percent of all deaths.

cisely, it is usually possible to estimate the logarithm within ±0.5. (A war of magnitude 6 ± 0.5 could have anywhere from 316,228 to 3,162,278 deaths.) But the use of logarithmic magnitudes has a psychological benefit as well: one can survey the entire spectrum of human violence on a single scale.

Random Violence

Richardson's war list (as refined by Wilkinson) includes 315 conflicts of magnitude 2.5 or greater (or, in other words, with at least about 300 deaths). It's no surprise that the two world wars of the twentieth century are at the top of this list; they are the only magnitude-7 conflicts known in human history. What *is* surprising is the extent to which the world wars dominate the overall death toll. Together they account for some 36 million deaths, which is about 60 percent of all the quarrel deaths in the 130-year period. The next-largest category is at the other end of the spectrum: the magnitude-0 events (quarrels in which 1 to 3 people died) were responsible for 9.7 million deaths. Thus the remainder of the 315 recorded wars, along with all the thousands of quarrels of intermediate size, produced less than a fourth of all the deaths.

The list of magnitude-6 wars also yields surprises, although of a different kind. Richardson identified seven of these conflicts, the smallest causing half a million deaths and the largest about two million. Clearly these are major upheavals in world history; you might think that every educated person could name most of them. Try it before you read on. The seven megadeath conflicts listed by Richardson are, in chronological order and using the names he adopted: the Taiping Rebellion (1851–64), the North American Civil War (1861–65), the Great War in La Plata (1865–70), the sequel to the Bolshevik Revolution (1918–20), the first Chinese-Communist

War (1927–36), the Spanish Civil War (1936–39), and the communal riots in the Indian Peninsula (1946–48).

Looking at the list of 315 wars as a time series, Richardson asked what patterns or regularities could be discerned. Is war becoming more frequent, or less? Is the typical magnitude increasing? Are there any periodicities in the record, or other tendencies for the events to form clusters?

A null hypothesis useful in addressing these questions suggests that wars are independent, random events, and on any given day there is always the same probability that war will break out. This hypothesis implies that the average number of new wars per year ought to obey a Poisson distribution, which describes how events tend to arrange themselves when each occurrence of an event is unlikely but there are many opportunities for an event to occur. The Poisson distribution is the law suitable for tabulating radioactive decays, cancer clusters, tornado touchdowns, Web-server hits, and, in a famous early ex-

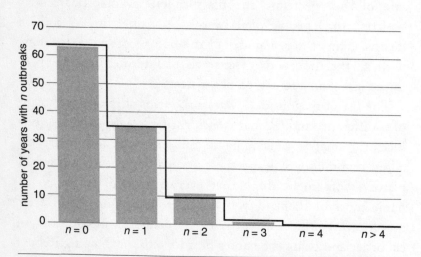

The frequency of outbreaks of war (gray bars) is closely modeled by the Poisson distribution (black line), suggesting that the onset of war is a random process. The data are for magnitude-4 wars over 110 years.

ample, deaths of cavalrymen by horse kicks. As applied to the statistics of deadly quarrels, the Poisson law says that if r is the average annual rate at which wars begin, then the probability of seeing n wars begin in any one year is $e^{-r}r^n/n!$, where e is the base of the natural logarithms (about 2.718) and $n!$ indicates the factorial of n—the product of all the integers from 1 through n. When r is less than 1, years with no onsets of war are the most likely, followed by years in which a single war begins; years with still more wars are less frequent.

The graph on the opposite page compares the Poisson distribution with Richardson's data for a group of magnitude-4 wars. There were 60 of these wars in 110 years, so that r in the formula above is equal to 0.545. The match between theory and observation is very close. Richardson performed a similar analysis of the dates on which wars ended—the "outbreaks of peace"—with the same result. Thus the data offer no reason to believe that wars are anything other than randomly distributed accidents.

Richardson also examined his data set for evidence of long-term trends in the incidence of war. Although certain patterns catch the eye when the data are plotted chronologically, Richardson concluded that the trends are not clear enough to rule out random fluctuations. "The collection as a whole does not indicate any trend towards more, nor towards fewer, fatal quarrels." He did find some slight hint of "contagion": the presence of an ongoing war may to some extent increase the probability of a new war starting.

Love Thy Neighbor

If the temporal dimension fails to explain much about war, what about spatial relations? Are neighboring countries less likely than average to wind up fighting each other, or more

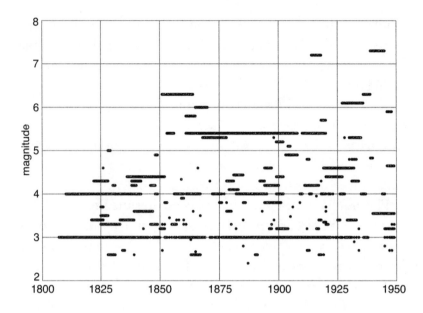

The distribution of wars over the decades reveals no clear pattern in Richardson's catalog of 315 conflicts. Although the eye may detect an apparent increase in high-magnitude wars, Richardson's statistical tests failed to confirm this trend.

likely? Either hypothesis seems defensible. Close neighbors often have interests in common and so might be expected to become allies rather than enemies. On the other hand, neighbors could also be rivals contending for a share of the same resources—or maybe the people next door are just plain annoying. The existence of civil wars argues that living together is no guarantee of amity. (And at the low end of the magnitude scale, people often murder their own kin.)

Richardson's approach to these questions had a topological flavor. Instead of measuring the distance between countries, he merely asked whether or not they share a boundary. Then, in later studies, he refined this notion by trying to measure the length of the common boundary—which led to a fascinating digression. Working with maps at various scales, Richardson paced off the lengths of boundaries and coastlines with dividers, and realized that the result depends on the setting of the dividers, or in other words on the unit of measurement. A coastline that measures one hundred steps of ten millimeters each will not necessarily measure one thousand steps of one millimeter each; the finer scale is likely to yield a longer measurement, because the smaller units more closely follow the zigzag path of the coast. This result appeared in a somewhat out-of-the-way publication; when Benoit Mandelbrot came across it by chance, Richardson's observation became one of the ideas that inspired Mandelbrot's theory of fractals.

During the period covered by Richardson's study, there were about sixty stable nations and empires (the empires being counted for this purpose as single entities). The mean number of neighbors for these states was about six (and Richardson offered an elegant geometric argument, based on Leonhard Euler's relation among the vertices, edges, and faces of a polyhedron, that the number must be approximately six for any plausible arrangement of nations). Hence if warring nations were to choose their foes entirely at random, there would be about a 10 percent chance that any pair of belligerents would turn out to be neighbors. The actual proportion of warring neighbors is far higher. Of ninety-four international wars with just two participants, Richardson found only twelve cases in which the two combatants had no shared boundary, suggesting that war is mostly a neighborhood affair.

But extending this conclusion to larger and wider wars proved difficult, mainly because the "great powers" are effectively everyone's neighbors. Richardson was best able to fit the data with a rather complex model assigning different probabilities to conflicts between two great powers, between a great power and a smaller state, and between two lesser nations. But rigging up a model with three parameters for such a small data set is not very satisfying. Furthermore, Richardson concluded that "chaos" was still the predominant factor in explaining the world's larger wars: the same element of randomness seen in the time-series analysis is at work here, though "restricted by geography and modified by infectiousness."

What about other causative factors—social, economic, cultural? While compiling his war list, Richardson noted the various items that historians mentioned as possible irritants or pacifying influences, and then he looked for correlations between these factors and the prevalence of war. The results were almost uniformly disappointing. Richardson's own suppositions about the importance of arms races were not confirmed; he found evidence of a preparatory arms race in only 13 out of 315 cases. Richardson was also a proponent of the artificial language Esperanto, but his hope that a common language would reduce the chance of conflict failed to find support in the data. Economic indicators were equally unhelpful: the statistics ratify neither the idea that war is a struggle between the rich and the poor nor the view that commerce between nations creates bonds that prevent war.

The one social factor that does have some detectable correlation with war is religion. In the Richardson data set, nations that differ in religion are more likely to fight than those that share the same religion. Moreover, some sects seem generally to be more bellicose (Christian nations participated in a disproportionate number of conflicts). But these effects are not large.

Mere Anarchy Loosed upon the World

The residuum of all these non-causes of war is mere randomness—the notion that warring nations bang against one another with no more plan or principle than molecules in an over-heated gas. In this respect, Richardson's data suggest that wars are like hurricanes or earthquakes: We can't know in advance when or where a specific event will strike, but we do know how many to expect in the long run. We can compute the number of victims; we just can't say who they'll be.

This view of wars as random catastrophes is not a comforting thought. It seems to leave us no control over our own destiny, nor any room for individual virtue or villainy. If wars just happen, who's to blame? But this is a misreading of Richardson's findings. Statistical "laws" are not rules that govern the behavior either of nations or of individuals; they merely describe that behavior in the aggregate. A murderer might offer the defense that the crime rate is a known quantity, and so *someone* has to keep it up, but that plea is not likely to earn the sympathy of a jury. Conscience and personal responsibility are in no way diminished by taking a statistical view of war.

What *is* depressing is that the data suggest no clear plan of action for those who want to reduce the prevalence of violence. Richardson himself was disappointed that his studies pointed to no obvious remedy. Perhaps he was expecting too much. A retired physicist reading the *Encyclopaedia Britannica* can do just so much toward securing world peace. But with larger and more detailed data sets, and more powerful statistical machinery, some useful lessons might emerge.

There is now a whole community of people working to gather war data, many of whom trace their intellectual heritage back to Richardson and Quincy Wright. The largest such undertaking is the Correlates of War project, begun in the 1960s by J. David Singer of the University of Michigan. The COW catalogs, like Richardson's, begin in the post-Napoleonic period, but they have been brought up close to the present day and now list thousands of militarized disputes.

Peter Brecke of the Georgia Institute of Technology has begun another data collection. His catalog extends down to magnitude 1.5 (about thirty deaths) and covers a much longer span of time, back as far as A.D. 1400. The catalog is approaching completion for five of twelve global regions and includes more than three thousand conflicts. The most intriguing finding so far is a dramatic hundred-year lull in the eighteenth century.

Even if Richardson's limited data were all we had to go on, one clear policy imperative emerges: at all costs avoid the clash of the titans. However painful a series of brushfire wars may seem to the participants, it is the great global conflagrations that threaten us most. As noted above, the two magnitude-7 wars of the twentieth century were responsible for three-fifths of all the deaths that Richardson recorded. We now have it in our power to have a magnitude-8 or -9 war. In the aftermath of such an event, no one would say that war is demographically irrelevant. After a war of magnitude 9.8, no one would say anything at all.

AFTERTHOUGHTS

The essay above was written a few months after the disasters of September 11, 2001, as NATO forces were invading Afghanistan and rhetoric was building for war with Iraq. As I write *these* words, five years later, both of those conflicts continue, and indeed they have lately grown more intense and lethal. Estimates of the total number killed vary over a wide range, but it seems likely the Iraq war has reached magnitude 5 by Richardson's way of counting (at least 31,623 deaths, but fewer than 316,228). Elsewhere in the world, there is no shortage of other deadly quarrels: Sudan, Somalia, the Great Lakes region of Africa, Israel and Palestine, Lebanon, Sri Lanka, Indonesia, Kashmir, Colombia. Can Richardson's dispassionate mathematical view of these events really help us to understand them—much less to reduce their frequency or severity? I still think so, but I have moments of doubt. For that matter, so did Richardson.

One frequent objection to the statistical approach is that wars are not at all random or unpredictable events. They have deep-seated causes: greed, aggression, vengeance, racial hatred, religious intolerance. Sometimes nobler motives are invoked as well: the urge to liberate the oppressed, redress an injustice, come to the aid of a friend. And many wars are blamed on the behavior of individuals: bullies, tyrants, zealots, fools, "evildoers who hate freedom."

It's all true: nations do not pick fights at random, and often they have good reason for taking up arms. But to say that war can be modeled by a random process is not to say that leaders are making decisions by flipping a coin (though perhaps we'd be better off if they were). Every war is unique, but over decades or centuries all the peculiar causes and circumstances seem to average out, and we're left with a persistent statistical pattern that seems quite insensitive to other historical trends. Empires come and go; so do ideologies and even religions, but war marches on through it all. An analogy may help explain this point. Every automobile accident has its own set of causes: the icy bridge, the sleepy driver, glare from the setting sun. But without knowing anything about these details, we can predict the total number of accidents per year quite accurately.

Several readers pointed out that Richardson neglected an important variable when he examined the various factors that might affect the incidence of war. He considered language and trade relations and religion but not the nature of a country's political system or form of government. This omission is particularly significant in view of the widely held belief that no two democratic states have ever gone to war against each other. This assertion can be traced back to a 1983 article by Michael W. Doyle of Columbia University, who stated it rather carefully: "Constitutionally secure liberal states have yet to engage in war with one another." A number of other scholars have since examined the issue, among them Zeev Maoz, Nasrin Abdolali, Christopher Layne, Bruce Russett, and David E. Spiro.

Is it factually true that democracies have never gone to war with each other? Everything hinges on how you define "democracy." If Britain and the United States were democracies early in the nineteenth century, then the War of 1812 would be a counterexample. Both nations had elected governments then, but perhaps they were not yet sufficiently representative to be counted as truly democratic. On the other hand, if you set the standard too high, then the proposition that democracies don't fight among themselves becomes a vacuous truth, because there have been precious few democracies until very recent years. For example, if something close to universal suffrage is the criterion for democracy, then the United States was not a democratic nation until at least 1920 (when women gained the right to vote) or perhaps 1965, with the passage of the Voting Rights Act, which promises African-Americans access to the polls.

In sorting out the connections between form of government and militarism, one must also address the question of which is the cause and which the effect. War and the prelude to war put stress on democratic institutions. By the time a country is embroiled in military violence, it may look a lot less democratic than it did in peacetime. In other words, wars between democracies may be absent or rare not because democracy inhibits war but because war undermines democracy.

Setting aside these doubts and quibbles, however, I have to agree that Richardson's neglect of the issue is curious and disappointing. It certainly merits more attention now.

What worries me most about the quantitative approach to understanding war is the difficulty of gathering quantitative data. Richardson's method demands reasonably accurate (or at least consistent) records of the numbers killed in each conflict, but trustworthy counts are hard to come by. We keep careful tallies of how many people die on the highways, but no one seems to know the true death toll on the battlefield. Even for recent and well-studied wars, the estimates vary over an alarmingly wide range. For the Vietnam War (considering only the period of American involvement) the Correlates of War project estimates total deaths at just over a million, but the *Encyclopaedia Britannica* offers a figure of more than two million. How can we measure the effects of war if we can't even count the dead to the nearest million?

Or maybe it doesn't matter so much. Would a war be only half as bad if half as many people died?

CHAPTER 6

Dividing the Continent

It was the fourth day of a meandering coast-to-coast road trip. We were climbing through the Centennial Mountains along the Idaho-Montana border in an overloaded Toyota with a U-Haul luggage pod on the roof. As we crested Monida Pass, a sign at the roadside announced: "Continental Divide, Elevation 6,823 Feet." "Well," quipped my traveling companion, "I guess it's all downhill from here."

For miles afterward—as we climbed still higher hills and crossed the divide twice more—I pondered that remark. The Great Divide is the spine of the continent: rain falling on one side trickles into the Pacific, and on the other side into the Atlantic. The concept is simple enough, but I kept wondering how we would have known we were crossing the divide if the highway department had not put up those helpful signs. The divide is not necessarily the high point of a cross-country journey, so what distinguishes it, geometrically or topologically? That morning in Idaho, it seemed even more enigmatic than other lines that people draw on the landscape. For example, a contour—a line connecting points of equal elevation—is something you could trace out by carrying around an altimeter, but there is no portable instrument that would help you find and follow the Continental Divide.

The long drive home offered us ample opportunity to noodle away at this puzzle. Since I'm a computer-dependent person, my instinct was to address the question in algorithmic terms; I would know that I understood the answer when I could write a program to identify the divide. Out on the road, however, I could not put such a program to the test. I am also a library-dependent person, but the urge to go find out what others had to say was also frustrated. And thus for a week or so I had no choice but to actually think about the problem.

Dividing the Ant Farm

In a two-dimensional world, it's easy to find a continental divide, if it exists. Think of an ant farm: a thin layer of soil sandwiched between two upright panes of glass. An ant walking along the surface of the soil from west to east will trace a one-dimensional profile, a graph of elevation as a function of longitude. If the profile has just one peak—that is, if the ant climbs steadily to some maximum elevation and thereafter descends continuously—then obviously that unique peak is the divide. ("It's all downhill from here," the ant might well say.) If there are multiple peaks with valleys between them, the highest of the summits must be the divide.

Some pathological possibilities could spoil this easy analysis. The ant-farm profile could have several tallest peaks, all at exactly the same height, or a plateau might form a continuous line of highest points. In these cases there is no unique continental divide. But such landforms are unlikely. Ignoring them, the algorithm for finding the ant-farm divide is straightforward: just look for the highest point.

When we leave behind the ant farm and consider a two-dimensional surface embedded in three-dimensional space, the divide problem gets more interesting. In particular, the find-the-maximum algorithm no longer works. Just try it for the case of North America! When you search out the highest point in the lower forty-eight states, you find yourself atop Mount Whitney, in California, elevation 14,500 feet. But Mount Whitney is nowhere near the Continental Divide, and all the water that falls on its flanks winds up in the Pacific, none in the Atlantic. (Indeed, much of it reaches the Pacific via the municipal water mains and the sewer system of Los Angeles.)

Thinking about this phenomenon on a larger scale raises doubts about the whole concept of a continental divide. Just as runoff can sneak around Mount Whitney, it can also find a path around the entire American cordillera, which doesn't really separate the Pacific from the Atlantic. After all, you can get from New York to San Francisco without climbing even the smallest hill: there is a sea-level route, around Cape Horn. From a topological point of view, a continental divide can exist only if a continent girdles the planet, so that the divide is a closed curve, with an inside and an outside.

Perhaps the best answer to this complaint is that the idea of a great divide belongs to the field of topography, not topology. Insisting on mathematical rigor is not necessarily helpful. In any case, we can rescue the concept of the divide, at the cost of making it somewhat artificial. The key step is to cut away a rectangular section of the earth's crust, corresponding roughly to the lower forty-eight states, and put it in a high-walled glass box—a terrarium, not so different from the ant farm. Now a continental divide is either a closed curve that lies entirely inside the box or a continuous line whose endpoints are anchored to the glass walls. With this definition, the divide truly does divide the territory into separate regions.

Think Globally, Classify Locally

How should the terrain inside the glass box be represented mathematically? One elegant idea is to make the earth's surface the graph of a continuous function $h(x,y)$, which defines a height h for every combination of x and y coordinates. (I assume the spherical surface is projected onto a plane. Also, it's necessary to smooth out cliffs and overhangs, so that every x,y pair yields a unique h.) The main advantage of this scheme is that you can find maxima and minima of the height function by standard methods of calculus—taking the derivative of h and looking for places where its value is zero. The disadvantage is that an equation for the surface of all of North America is likely to be quite unwieldy.

A more practical alternative is a discrete model, defining the elevation of the surface only at a finite number of points arranged in a grid. To keep things simple, let the grid be rectangular, formed by the intersections of evenly spaced north-south and east-west lines; it's like a lumpy fishnet draped over the hills and valleys of the continent. Each grid point connects only to its four nearest neighbors, so that water on the model landscape flows only in the four cardinal directions. With this model of the terrain, the task of a continental-divide algorithm can be stated more concretely. For each grid point, the algorithm must answer the questions: Does this point lie on the divide? Does it shed water into two or more basins that do not communicate with one another? Or, as with Mount Whitney, does all the water eventually drain into one ocean?

As we motored on into Montana, I tried to come up with a quick-and-easy algorithm to answer these questions. My first thought was to make the most of local information about the immediate neighborhood surrounding each grid point. I reasoned that the divide ought to run mainly along ridges, and ridges can be recognized by a distinctive pattern of higher and lower neighbors.

But analyzing those neighborhoods proved messy. Each of a point's 4 neighbors can be above, below, or level with the central point, and so there are $3^4 = 81$ possible configurations. That was too many cases to keep in mind while cruising down the interstate, so I decided to simplify by pretending that no two adjacent grid points are ever at exactly the same height. (This ruse is not as unrealistic as it might seem; if you could measure elevations with infinite precision, the probability of finding two identical values would be zero.)

When all neighbors must be either higher or lower, there are sixteen local configurations, and they can be further consolidated into just six classes. A *peak* is a point higher than all four of its neighbors, and a *pit* is lower than all of its neighbors. A point with exactly three lower neighbors lies on a *ridgeline*. The opposite case of three higher neighbors describes points along a valley bottom—a line known to topographers and crossword solvers as a *thalweg*. Finally, the points with equal numbers of higher and lower neighbors fall into two subclasses. If you stand on the central point and turn through 360 degrees to survey the neighbors, you might see them in a sequence such as above-above-below-below; a point of this kind lies on a *slope*. If the neighbors alternate, as in the sequence above-below-above-below, then the point is a *saddle* or *pass*. A saddle is special: it is the only kind of point that lies both on a ridgeline and on a thalweg.

Somewhere between Butte and Bozeman I began to doubt that this local classification was going to yield a useful algorithm for identifying the Continental Divide. It doesn't even help much in the one-dimensional ant-farm case (where the grid is merely a line and each point has only two neighbors instead of four). A divide in the ant farm always lies at a peak, but the local properties of peaks will not tell you which of them is the divide. Finding the highest of the peaks requires global information; you have to compare points throughout the entire grid.

For the two-dimensional surface, the situation is even stickier. Not only is local information unable to identify the divide, but in addition there is no simple *global* property that will settle the issue. In the ant farm you can compare each point with all the other points, looking for a maximum elevation. That doesn't work on the fishnet surface. Instead, you need to consider multiple pathways through the array of points.

But if local configurations can't solve the divide problem, maybe they can at least rule out lots of points that might otherwise be candidates for the divide. For example, it seems beyond question that a pit cannot form part of the divide, and so all pits can be crossed off the list. But as we rolled on beyond Bozeman toward Billings, I gradually realized that no other kinds of points could be excluded. Peaks and ridges and saddles are clearly allowed on the divide. It might seem that slopes and thalwegs would be ineligible, but this is not so. Think of a river delta, where streams diverge and bifurcate. If such a delta were to form at the outlet of a high alpine valley right on the Continental Divide, with channels flowing down either side, then the entire area of the upstream valley would have to be considered part of the divide, including the slopes and the thalweg.

Admittedly, a river delta in the highland headwaters is a pretty unlikely landform, but algorithms are supposed to cope with even the oddest cases. And natural landscapes do offer oddities. Some maps show the Great Divide itself dividing in Wyoming, where it envelops a high, craterlike basin. If those maps are correct, rain falling into the crater flows into neither ocean. Even stranger are some *un*natural landscapes. At the Big Thompson Project in Colorado a tunnel carries water across the Great Divide, or rather under it, or under the ridge where the divide ought to run. Such meddling with drainage patterns makes nonsense of the very concept of a divide. Suppose it were done on a larger scale: the Ohio River might be diverted through a tunnel under the Rockies and dumped into the Colorado. Then Pittsburgh would become part of the Pacific basin, whereas Denver would remain in the Atlantic.

Divide and Conquer

Over the next two hundred miles, as we followed the eastbound Yellowstone River (waters destined to reach the Atlantic via the Missouri, the Mississippi, and the Gulf of Mexico), I had several more bright ideas that proved faulty. For example, I thought I saw a way to extend the ant-farm algorithm to a higher-dimensional world. Having put the continent in a terrarium and draped it with a fishnet grid, I would first survey elevations along the south wall of the terrarium, and find the maximum. Then I'd set out from this peak, moving to the highest neighboring point, then to the highest neighbor of *that* point, and so on. I would stop when I came to another wall or when the path looped back on itself. I briefly believed that this procedure would trace out the divide. I had forgotten that the high point on the southern boundary is not necessarily on the divide, and so the algorithm might go wrong from the first step.

An attempt to patch up this idea led to another strategy, which I still find appealing even though it seems to be a dead end. Suppose you know the two points where a divide touches the perimeter of a square region, but the path through the interior of the square is entirely hidden. If you could cut the square into four smaller tiles, and determine where the divide crosses each of their sides (if at all), then you would begin to have a rough vision of its route. Quartering each of these squares yields sixteen more tiles, and then sixty-four. Once the squares get small enough, it is enough merely to know which sides of a square are crossed by the path of the divide, without trying to measure the exact position of the intercept. This is a classic divide-and-conquer algorithm, with a hint of deep recursive magic. Unfortunately, the magic is an illusion. The algorithm's initial supposition—that we know where the divide enters and exits each square—is unfounded.

Yet another idea also exploits the distinctive topology of the divide—the fact that it is either a closed curve or a curve with endpoints anchored to the boundary. Here's the plan. Working from local neighborhood information, identify all the grid points classified as peaks and ridgelines, and paint them red. The labeled peaks and ridges will form a dense network, which probably includes most or all of the divide, but it will extend to many other points as well. To be able to see the divide clearly, you need to prune away the underbrush. Topology suggests a promising way to do it. The premise is that all the peaks and ridges except those on the divide must have at least one free end—a dangling terminal point like the tip of a tree limb. You can find these free ends by scanning the grid for red points that have only one red neighbor. Removing this endpoint exposes a new terminal point, which is then subject to removal in the same way. If you repeat the scan until no more single-neighbor red sites can be found, only the divide will remain labeled—or so I thought.

In some instances this procedure actually works, but it has serious weaknesses. In the first place, as noted above, the divide does not have to lie entirely on peaks and ridges, and so the first stage of the algorithm may fail to label some points it should. The second stage can also introduce errors. Suppose two closed loops of the divide are connected by a ridge that is not properly part of the divide system (water on both sides of this ridge flows to the same basin). Because this extraneous section of ridge has no free end, it cannot be removed by the pruning procedure.

Long Division

However tricky the divide problem may prove to be, a correct algorithm surely does exist, since nature somehow finds a solution. If all else fails, one can emulate the natural algorithm. The idea is to let raindrops fall on each grid point, and then follow the runoff as it drains toward lower elevations. The most thorough version of this algorithm pursues every downhill path. That is, if a point has three lower neighbors, then the algorithm follows droplets that roll along each of the three downhill links. The flow stops when the droplet reaches a pit and has nowhere more to go.

Tracing such paths for all points on the continent should identify the Great Divide. The divide is just the set of points from which droplets reach both the Atlantic and the Pacific basins. (This is, after all, the definition of the divide.)

The rainfall algorithm works, at least for small test cases, but it is fabulously inefficient. An average point on the model landscape has two downhill neighbors, which means the number of paths to be explored doubles at every step. If a typical path is just twenty steps long, the algorithm will have to map a million paths for each point.

This exponential explosion of pathways should not be necessary. Real water droplets don't explore all possible routes to the sea; for the most part, they stick to the path of steepest descent. An algorithm can do the same, which makes the computational burden much lighter. But there are other problems. The divide has to be defined somewhat differently—as a path threading between grid points rather than a connected series of points.

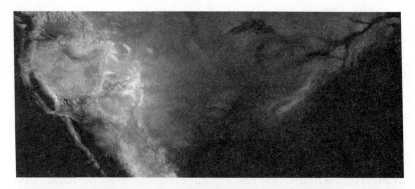

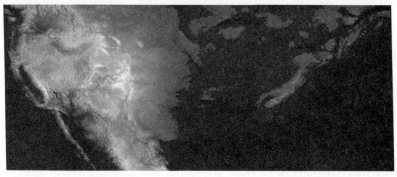

The global-warming algorithm finds the Great Divide by raising sea level until the Atlantic and Pacific meet (but don't mix). Starting from present sea level (upper left), the sea begins to encroach on the continent, flooding most of the East (lower left). As the level continues rising, the two oceans first meet at a point near Deming, New Mexico; this

And there is the lake-bottom problem: a path of steepest descent seldom descends continuously from the divide all the way to the ocean. It's *not* all downhill from here.

Somewhere in North Dakota or Minnesota—near another divide that separates Hudson Bay from the Gulf of Mexico—I finally began to settle on an idea that might be called the global-warming algorithm. It works like this: Given North America in a terrarium, start raising the sea level, and keep the

lowest section of the Continental Divide appears on the map as a short
segment of bright white (upper right). When the sea has inundated the
entire continent, the full path of the divide is revealed (lower right).
Elevation data for the model are from the National Oceanic and
Atmospheric Administration.

floods coming until the Atlantic and the Pacific just touch. At
this moment you have identified one point—namely the lowest
point—on the Great Divide. Now continue adding water, but
as the sea level rises further, don't allow the two oceans to mix;
raise a barrier between them, just high enough to keep them
separate. (This would be a difficult trick in a physical model,
and it's none too easy even in a computer simulation.) Note the
succession of points where east meets west, and mark them

down as elements of the divide. When the last such point is submerged, you have succeeded in dividing the continent.

Describing this process in terms of water filling a basin tends to conceal some of the nitty-gritty computational details. Real water is very good at flooding; it just "knows" how to do it and never makes a mistake. Simulated water, on the other hand, must meticulously plot its every move. To raise the level one foot, you have to check every point adjacent to the current waterline and decide which points will be newly submerged. Then you have to look at the neighbors of these selected points, and at their neighbors, and so on. There's the potential for another exponential explosion here, although with realistic landscapes it doesn't seem to happen.

When I finally got a chance to write a program for this process, I found that the algorithm is exquisitely sensitive to the order of operations. Consider the situation just as the Pacific is about to reach the lowest point on the divide. If the Atlantic has not been raised in synchrony, then the Pacific waters will pour over the saddle point and flood part of the eastern basin, shifting the divide to an incorrect position. With sufficient attention to the details, however, the method does seem to work.

I tested the program with a digitized elevation map created by the National Oceanic and Atmospheric Administration. For the section of North America I included in my glass-walled terrarium, the data set gave the elevation of the land surface in meters above or below sea level at 217,021 points arranged in a rectangular grid.

Landscape Images

Back home again, and plugged into libraries as well as computers, I was not surprised to learn that others had gone before me in thinking about the nature of watersheds and divides. But I was surprised to learn just who my predecessors were.

Two of the best publications on the subject are short papers by Arthur Cayley (a founder of topology and graph theory) and James Clerk Maxwell (the author of the electromagnetic theory of light). Cayley and Maxwell do not focus on continental divides—perhaps the concept is not an obvious one for residents of an island nation—but their analysis of landforms in general clarifies aspects of the divide problem. They emphasize peaks, pits, and saddles as the keys to delineating the fundamental regions of a landscape.

Much as Leonhard Euler gave a formula for the number of faces, edges, and vertices in a polyhedron, Maxwell relates the number of topographic peaks, pits, and saddles on a surface. For a landscape wrapped around the surface of a sphere, the formula is $p + q - s = 2$, where p is the number of peaks, q the number of pits, and s the number of saddles. Maxwell also outlines a procedure for dividing the landscape into watershed regions. Whereas my own methods progressed either down from peaks or up from pits, Maxwell argues that the right way to do it is to start in the middle—at a saddle—and proceed toward both peaks and pits along lines of steepest ascent or descent.

The more recent literature on divides and watersheds held another surprise. I had expected to find work by geographers and cartographers, and indeed they have written extensively on the subject. But there is also a body of work by students of image analysis and artificial vision. The connection is clear once it's pointed out. Just as a topographic map can be presented as an image in which elevation is encoded in brightness, so a digitized image can be interpreted as a surface, where the altitude of each point encodes the color or shade of gray at the corresponding position in the image. Finding watersheds in such a surface is a useful approach to identifying objects in the image.

The idea of using watersheds for image analysis was first proposed at the School of Mines in Paris in the 1970s. (The images that needed analyzing were micrographs of ore samples.) Workers there have continued to refine the method. The version of the algorithm I have found most helpful was devised by Luc Vincent and Pierre Soille when they were both working at the School of Mines.

The Vincent-Soille algorithm is related to what I have dubbed the global-warming method, but with a number of enhancements. One remarkably simple device greatly reduces the computational effort. In addition to storing the array of points that represents the landscape, Vincent and Soille keep a list of the same points sorted in order of increasing elevation. With this list in hand, there is no need to search for points that will be submerged each time the water level rises; you can simply cross them off in order.

As it happens, one contemplated application of the watershed algorithm in image processing is the old dream of a car that drives itself. The "watersheds" detected in a video image might be the edges of a roadway. So perhaps the next time I cross the Continental Divide, I'll be able to pay more attention. I won't have to keep my hands on the wheel.

AFTERTHOUGHTS

The road trip narrated in this essay took place in the summer of 2000, and the article was first published in *American Scientist* at the end of that year. In telling the story, I wanted to focus not so much on the solution to the watershed problem as on the process by which people go about finding or inventing solutions. Where do the ideas come from? How do we evaluate alternatives? How do we know when to stop? If you are searching for your lost car keys, they are always in the last place you look—because, of course, there's no reason to go on looking once they're found. The search for an ideal algorithm is more open-ended. In looking for a method to delineate the Continental Divide, I stopped when I found an algorithm that worked, but there's no compelling reason to believe that solution is the best one.

Several readers of the essay proposed simpler algorithms, mostly hill-climbing routines. The idea is to start by finding all the low points of the landscape—the ocean basins and also any other local minima of the elevation map (in essence, the points I have designated as pits). Mark each low point with a unique color. Now visit each neighbor of each marked point; if the neighbor is at a higher elevation and is not already colored, apply the color of the lower point you've just come from. Stop whenever you reach a point that has no higher uncolored neighbors. At the end of this process, every point should be colored, and the map should be partitioned into regions that define separate catchment basins. The boundaries between differently colored regions are the watersheds, or divides.

Does this method work? It rests on the rationale that water flows only downhill, which seems like an unassailable truth. And indeed the algorithm does correctly identify watersheds— but only for a waterless planet! The unstated assumption is that once water flows down to the bottom of a basin, it can never get out again. If there's enough water, however, the basin eventually fills up and overflows. Thus most basins are not the final destination of a raindrop.

Of the 217,021 points in my North American elevation map, 10,241 are classified as pits, and so the hill-climbing algorithm divides the continent into 10,241 basins, separated by a lacy network of ridgelines. A few of these basins really do deserve to be viewed as independent watersheds, with the same status as the oceans, because the water that reaches them never escapes except by evaporation. Great Salt Lake is the most prominent example. But most of the basins are lakes and ponds that *do* have an outlet. Water that flows into them ultimately flows out again, and thus properly belongs to some downstream catchment—almost always one of the oceans.

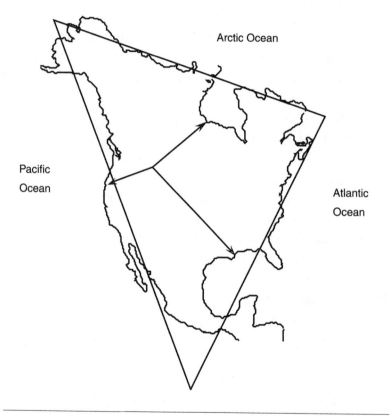

Stewart Rood suggests seeing the North American continent not as a rectangle with a line dividing east from west but as a triangle with three great drainage basins, leading to the Atlantic, the Pacific, and the Arctic oceans. (Map courtesy of Stewart Rood)

Perhaps the root of the problem is that the concept of a divide is easier to define mathematically than it is geographically. Alan P. Peterson wrote to point out that one South American divide is particularly vague. You can travel by boat up the Orinoco River and then, following a natural waterway called the Casiquiare Canal, continue on to the Rio Negro and then go down the Amazon. Where is the parting of the waters? You cross from one watershed to another without ever getting out of the boat.

Wayne Slattery called my attention to another distinctive topographic feature, this one on the North American divide. I had mentioned the theoretical possibility of a stream running along the divide and splitting into Atlantic and Pacific branches. That would be an unlikely landform, I said, but Slattery observes that such a place exists. In northwestern Wyoming, in Grand Teton National Park, Two Ocean Creek splits into Atlantic Creek and Pacific Creek, with the Continental Divide running right through the point of bifurcation.

Stewart Rood, a Canadian reader, suggests that my Left Coast–Right Coast view of North America is somewhat parochial. Seen from a greater height and a higher latitude, the outline of the continent is roughly triangular, with the Third Coast being that of the Arctic Ocean. The "hydrographic apex of the continent," Rood notes, is at Triple Divide Peak in Glacier National Park, Montana. Below the slopes of this mountain we again find an Atlantic Creek and a Pacific Creek, but in addition there is a stream named Hudson Bay Creek, which runs north and eventually joins the Nelson River, reaching the Arctic Ocean via Hudson Bay.

CHAPTER 7

On the Teeth of Wheels

For many years, the basic raw material of the computer industry was not silicon but brass. Calculators built before 1700 by Wilhelm Schickard, Blaise Pascal, and Gottfried Wilhelm Leibniz were all based on the meshing of metal gears. Later, Charles Babbage conceived elaborate fantasies of gear work for his calculating engines. Later still, Vannevar Bush put gears and other rotating parts at the heart of his differential analyzer. And all of these inventors were foreshadowed by anonymous artisans in the city of Rhodes in the first century B.C., who assembled more than thirty gears in a remarkable calendrical computer known as the Antikythera mechanism.

These examples testify to the importance of gears in the history of computing. Less obvious is the importance of computing in the history of gears. I was ignorant of the connection myself until a few years ago, when I went looking in the library for a work on number theory and found myself making a detour into the engineering shelves. I learned there that the designers of gear trains have not merely borrowed ideas from mathematics but have also developed some of those ideas on their own and lent them back to the mathematicians. Mechanical engineers doubtless know all about this two-way traffic between math and mechanism, but others may find the computational roots of gear design as surprising as I did.

The Stern-Brocot Tree

The story began when I met a young mathematician named Divakar Viswanath, now at the University of Michigan, who had devised a "randomized" version of the Fibonacci numbers. The original Fibonacci numbers were invented sometime around the year 1200 by the Italian merchant and mathematician Leonardo of Pisa, who had the nickname Fibonacci. The sequence of numbers begins 1, 1, 2, 3, 5, 8, 13, 21, . . . , where the basic rule is to form each new term of the series by adding the two previous terms. Viswanath's series is similar but includes an element of randomness: instead of always adding two terms, you either add or subtract, making the choice at random in each case. You might guess that random additions and subtractions would tend to cancel each other out, but Viswanath proved that the numbers grow steadily in absolute value.

In explaining his work on the randomized Fibonacci sequence, Viswanath introduced me to an object from number theory called the Stern-Brocot tree. The leaves and branches of this mathematical tree are rational numbers—ratios of integers, such as $1/3$ or $3/2$. The tree is constructed as follows: Take any two rational numbers, a/b and c/d, and insert between them a third value, called the mediant, equal to $(a + c)/(b + d)$. For example, if you start with $2/3$ and $3/4$, the mediant is $(2 + 3)/(3 + 4)$, or $5/7$. Now, with three numbers in hand, construct mediants between the first and the second and between the second and the third, so that the next level of the tree has five members. The process can continue indefinitely. Note that on each level the numbers are always in order.

The canonical version of the Stern-Brocot tree starts with the numbers $0/1$ and $1/0$. (The second of these "numbers" is admittedly peculiar; somebody has said it is "infinity reduced to lowest terms.") With these starting values, the second level of the tree consists of $0/1$, $1/1$, and $1/0$, and the third level becomes $0/1$, $1/2$, $1/1$, $2/1$, and $1/0$. The remarkable thing about the tree is that every rational number appears somewhere among the leaves and branches, but no number appears more than once.

When I described the Stern-Brocot tree in an *American Scientist* column, I mentioned that it is named for "the mathematician Moriz Stern and the watchmaker Achille Brocot." Now I must make a confession. Although Viswanath's paper cited the works of Stern and Brocot, I did not look up those references. At the time, I excused this lapse of diligence on the grounds that Viswanath himself gave a lucid account of the tree's con-

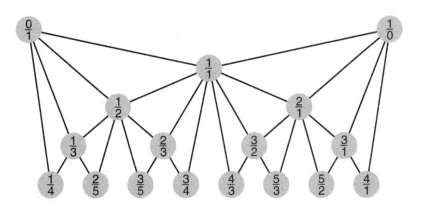

The Stern-Brocot tree includes every rational number once and only once somewhere among its leaves and branches. Each entry is formed by adding the numerators and the denominators of the left and right neighbors. Thus $1/3$ and $1/2$ yield $(1 + 1)/(3 + 2)$, which is equal to $2/5$.

struction, and I also had an excellent secondary source, namely the textbook *Concrete Mathematics*, by Ronald L. Graham, Donald E. Knuth, and Oren Patashnik. I knew what I needed to know of the tree without tracking down two obscure nineteenth-century papers written in German and French. Or I thought I knew. In fact I didn't know what I was missing.

The Professor and the Watchmaker

Thus the situation might have remained but for the prompting of a reader and friend, Horacio A. Caruso of La Plata, Argentina. Caruso and his young colleague Sebastián M. Marotta were sufficiently interested in the randomized Fibonacci phenomenon to undertake investigations of their own. For example, they extended Viswanath's idea to complex numbers, creating an intriguing series of fractal images. When Caruso inquired about the works of Stern and Brocot, I was belatedly inspired to go look them up.

Stern's paper was not hard to find. Moritz Abraham Stern (my earlier spelling Moriz was erroneous) was a prominent figure in the mathematical world of his day, a colleague of Carl Friedrich Gauss who succeeded Gauss as Ordinary Professor of Mathematics at the University of Göttingen. According to Peter Pulzer, Stern was "the first Jew to be appointed to a full professorship at a German university without first converting to Christianity."

Stern's paper appeared in the *Journal für die reine und angewandte Mathematik*, also known as *Crelle's Journal*, after its first editor, August Leopold Crelle. When the journal was founded in 1826, its title reflected the growing division in mathematics between *reine* and *angewandte*—"pure" and "applied." Stern's paper, "On a Number-Theoretical Function," is of the pure persuasion. He discusses several versions of the procedure for forming mediants and relates the sequence of mediants to other ways of constructing the set of rational numbers, such as continued fractions. Nowhere does he hint that his number-theoretical function might be of use to the makers of gears.

Finding Brocot's contribution was more challenging. His article was published in the *Revue chronométrique*, a French journal that commenced publication in 1855 and ceased in 1914. Only when I began searching for the *Revue* did it occur to me to wonder why a work on number theory was appearing in a clockmakers' journal.

None of the libraries within easy reach had the *Revue chronométrique*, but a catalog search at Duke University did return one hit for the term "Brocot." I was referred to a 1947 work by Henry Edward Merritt titled *Gear Trains: Including a Brocot Table of Decimal Equivalents and a Table of Factors of All Useful Numbers up to 200,000.* A Brocot table? Useful numbers? What was this all about? I walked from the mathematics library to the engineering library next door and soon had a worn blue volume in my hands. When I opened to the preface, I knew I would have to read the rest of the book. Merritt begins:

> Prefaces are not what they were. Who could resist the opening phrases of the editor's preface to the second edition of *Camus on the Teeth of Wheels*, published in 1836—
>
> "Always feeling annoyed at meeting with a long preface to a book, labouring as it were to beget a prepossession in favour of the author, and standing between the reader and the subject, like an impertinent porter, who detains a visitor at the gate, instead of giving him admission to the presence of the master, the editor will confine himself to five pages of preliminary remarks."

Indeed, who could resist? And, furthermore, what is this mysterious *Camus on the Teeth of Wheels?* The title would have been an apt one for the tormented existentialist Albert Camus, but it comes from the wrong century.

Counting Teeth

Reading on in Merritt's book, I soon learned why aspects of number theory have attracted the interest of gear makers. Somewhere in the works of a clock, you might have a shaft that turns once per minute; you want to design gears that will slow this motion to one revolution per day, which works out to a speed ratio of 1,440 to 1. The first law of gear work says that the speed of a gear is inversely proportional to the number of its teeth. Thus the most direct solution would be a driving gear

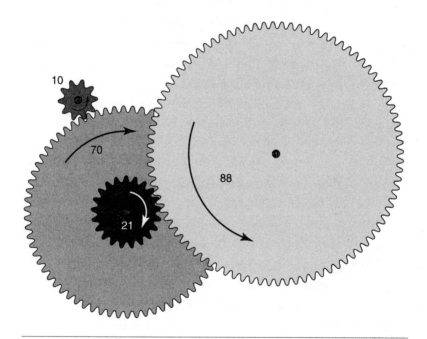

A compound gear train can achieve a large numerical gear ratio without requiring inconveniently large or small gears. The two pairs of gears here combine to yield an overall ratio of $^{10}\!/_{70} \times {}^{21}\!/_{88} = {}^{3}\!/_{88}$.

(a pinion) with just 1 tooth, meshing with a driven gear (or wheel) of 1,440 teeth. But a 1-tooth gear would be extremely awkward, and a 1,440-tooth gear is inconveniently large. You could solve the problem of the too-small pinion by multiplying both sides of the ratio by some convenient number, say 10. You would then have a pinion of 10 teeth, but of course the already-too-large wheel would be even larger, with 14,400 teeth.

The answer to this quandary is a compound gear train, where two or more pairs of mating gears progressively reduce the rotational speed. In a two-stage train, pinion a meshes with wheel A; then a second pinion, b, mounted on the same shaft as A, turns wheel B. The overall gear ratio is $a/A \times b/B$, and so you can choose any convenient values of a, A, b, and B that yield the correct product. For example, compound gears with the ratios $6/200$ and $5/216$ form the product $30/43{,}200$, which is equivalent to the required $1/1{,}440$. If wheels of 200 and 216 teeth are still too large, then a three-stage train with ratios of $6/72$, $6/60$, and $5/60$ would yield the same result.

The next question is: Where do numbers like $6/200$ and $5/216$ come from? It's easy to verify that they produce the correct gear ratio, but how do you find such numbers in the first place? The answer comes straight from number theory: factoring. In the ratio $30/43{,}200$, the numerator 30 has the prime factors $2 \times 3 \times 5$. Any combination of these numbers will work to produce the equivalent of a 30-tooth gear in a compound gear train: 6×5, 3×10, or 2×15. The denominator 43,200 breaks down into eleven prime factors: $2 \times 2 \times 2 \times 2 \times 2 \times 2 \times 3 \times 3 \times 3 \times 5 \times 5$, which can be partitioned into two groups in forty-one ways. A 200-tooth gear has the prime factors $2 \times 2 \times 2 \times 5 \times 5$, and the gear with 216 teeth accounts for the remaining factors $2 \times 2 \times 2 \times 3 \times 3 \times 3$.

This application of factoring explains the presence in Merritt's book of a "table of factors of all useful numbers up to

200,000." The "useful" numbers turn out to be those whose largest prime factor is no greater than 127, which Merritt suggests is a reasonable upper limit for the number of teeth on a gear. Number theorists have another word for the same concept: integers that have many small factors are called "smooth" numbers.

Does the need to factor numbers make the design of compound gear trains a hard computational problem? Factoring has an enigmatic status in computer science: for conventional computer hardware, the only known factoring algorithms are inefficient, and therefore slow in the worst case, but no one has proved that better algorithms cannot exist. For gear design, however, the issue of algorithmic intractability simply does not arise, because the factoring of smooth numbers is *always* easy. Even the crudest algorithm—a brute-force trial of all possible divisors—works quickly with numbers that have only small factors.

Gear Geeks

Converting minutes into days is a problem that gears can solve exactly, but what if the ratio of two speeds is π? Here the first law of gear work fails because it runs up against the zeroth law—that the number of teeth on a gear must be an integer. No ratio of integers can be equal to π. The best one can hope for is a good rational approximation. This is where Merritt's "Brocot table" enters the story, and it put me back on the trail of Brocot's original paper.

A visit to the New York Public Library proved tantalizing; I found several volumes of the *Revue chronométrique*, but not the volume I needed. On the other hand, the library was able to supply the enigmatic *Camus on the Teeth of Wheels*. Camus turned out to be Charles-Étienne-Louis Camus (1699–1768), author of *Cours de mathématique*, published in 1749. The section of this textbook dealing with gear work was extracted and translated into English by John Isaac Hawkins, a civil engineer, and published under the title *A Treatise on the Teeth of Wheels, Demonstrating the Best Forms Which Can Be Given to Them for the Purposes of Machinery; Such as Mill-work and Clock-work, and the Art of Finding Their Numbers.*

The first part of Camus's treatise deals with a geometrical rather than a number-theoretical question: What is the ideal shape for a gear tooth? This issue engaged the talents of mathematicians and other savants for generations. Robert Hooke, Thomas Young, Leonhard Euler, and Josiah Willard Gibbs all debated the merits of epicycloids and involutes. It's a fascinating problem, but my time was limited, and I turned to Part II of the treatise, where Camus takes up the numerical aspects of gear design.

For cases where an exact solution is possible, Camus explains the method of reducing a number to its prime factors and then partitioning the factors into as many groups as there are pairs of gears. He then turns to the task of approximating a ratio when the numbers have no convenient factorization. As an example he offers this problem: "To find the number of the teeth . . . of the wheels and pinions of a machine, which being moved by a pinion, placed on the hour wheel, shall cause a wheel to make a revolution in a mean year, supposed to consist of 365 days, 5 hours, 49 minutes." Multiplying out the days and hours yields a target ratio of $720/525,949$. The numerator of this fraction factors conveniently, but the denominator is a prime. Thus the aim is to find another fraction, as close as possible in value to $720/525,949$, but with both a numerator and a denominator that have small factors. Camus remarks: "In general this is done by repeated trials; but as this method is defective, we shall here propose another, by which the problem may be solved with more certainty."

But the next twenty pages, which set forth the method through worked examples, leave the impression that it's hardly much better than trial and error. Camus's procedure for finding ratios close to the target is a fairly arduous algebraic process, made worse by awkward and verbose notation. Furthermore, trial and error is still required, because there is no guarantee that a ratio generated by the method will be factorable. Camus reports seven failures before he hits on the ratio $^{196}/_{143,175}$, which can be factored as $^{4}/_{25} \times ^{7}/_{69} \times ^{7}/_{83}$. It was Brocot, a century later, who found a better way.

An Eminent Maker

I finally tracked down Brocot's memoir in the *Revue chronométrique* at the Mariners' Museum Library in Newport News, Virginia. (It's not such an unlikely place to go looking, given the close connections between seafaring and horology.) The library staff were able to help me find the article even though the citations I was working from turned out to have errors in both the volume number and the date.

A reference work on clockmakers lists Achille Brocot as "an eminent maker" and mentions his mechanical contrivances, such as the Brocot escapement, but it says nothing of contributions to mathematics. The article in the *Revue chronométrique* sticks to practicalities; the aim is to build a gear train, not to construct the infinity of rational numbers. And yet if theory is not emphasized, there is nonetheless something distinctly modern here. Brocot presents his method as an algorithm, albeit one adapted to pencil-and-paper methods rather than to programmable machinery.

As a pedagogical example, Brocot invents the problem of gearing a shaft that turns once in 23 minutes to another shaft that completes a revolution in 3 hours and 11 minutes, or in other words 191 minutes. Because 23 and 191 are both primes, gears with fewer teeth can only approximate the true ratio. To find the best such approximations, Brocot begins by noting that $^{191}/_{23}$ is greater than 8 but less than 9, so that the ratio must lie somewhere in the interval between 8 to 1 and 9 to 1. Accordingly, he writes in a row at the top of a sheet of paper:

$$8 \qquad 1 \qquad -7$$

Here the first two numbers represent the ratio 8 to 1, and the third number is the error associated with this initial approximation. The error is −7 because a ratio of $^{8}/_{1}$ is equal to $^{184}/_{23}$ rather than $^{191}/_{23}$, so that the slower wheel would complete its revolution 7 minutes early.

At the bottom of the same page, Brocot writes:

$$9 \qquad 1 \qquad +16$$

Again the first two numbers are an approximation to the ratio, and the third number is an error term, indicating that gears in a $^{9}/_{1}$ ratio will take 16 minutes too long to complete a revolution, since $^{9}/_{1}$ is equal to $^{207}/_{23}$.

Now the iterative part of the algorithm begins. Brocot adds the numbers in all three columns and writes the row of sums in the middle of the page:

$$17 \qquad 2 \qquad +9$$

This result is a new and more refined approximation. At a gear ratio of $^{17}/_{2}$, turning the faster shaft in 23 minutes causes the slower shaft to complete a revolution in 195.5 minutes, for an

error of 4.5 minutes. In the table, the error term of +9 is understood to represent the quantity $9/2$.

Brocot now has the choice of adding the middle row of numbers to those at the top of the page or to those at the bottom.
The mediant between top and middle is preferable because it
leads to a smaller error term. The result of the addition inserts
another row into the table:

25	3	+2

With this ratio the slower wheel turns in 191.67 minutes, for
an error of two-thirds of a minute.

Further approximations are constructed in the same way, always adding the latest entry in the table to whichever of its
neighbors reduces the error term. The method is deterministic,
with no need for guesswork or trial and error. And, like any
good algorithm, it is guaranteed to terminate eventually. The
end comes when the process converges on the original ratio
$191/23$, which necessarily has an error of zero. For Brocot's example, the final table looks like this:

8	1	−7
33	4	−5
58	7	−3
83	10	−1
191	23	0
108	13	+1
25	3	+2
17	2	+9
9	1	+16

Brocot's algorithm reveals that the closest approximations to
$191/23$ are ratios of $83/10$ (which runs a tenth of a minute fast) and

$^{108}/_{13}$ (a thirteenth of a minute slow). To do better requires a multistage gear train. Surprisingly, Brocot applies exactly the same algorithm to the design of such trains. He places one of the approximations at the top of the page and the exact ratio at the bottom. Then a series of additions produces successive ratios with smaller errors and larger numbers of teeth. For example, adding $^{191}/_{23}$ to $^{83}/_{10}$ yields $^{274}/_{33}$, and then adding $^{191}/_{23}$ yet again produces $^{465}/_{56}$:

83	10	−1
274	33	−1
465	56	−1
656	79	−1
.	.	.
.	.	.
.	.	.
191	23	0

If one of these combinations has a numerator and a denominator that can both be factored conveniently, the factors will offer the key to a superior approximation. But now trial and error *does* enter the process, because the only way to learn which numbers have small factors is to try factoring all of them. In this case, the third entry in the table can be factored as $^5/_7 \times {}^{93}/_8$, yielding a train of four gears that approaches the correct speed to within a fifty-sixth of a minute.

Brocot's algorithm can be employed as needed to find approximations to any given ratio, but Brocot also recognized that all the computation could be done beforehand and the results compiled into a table. This is the Brocot table included in Merritt's book; it is essentially a list of all fractions with numerator and denominator no greater than 100, ordered according to magnitude.

Shifting Gears

Just as Stern mentions no practical applications of his number-theoretical function, Brocot gives little attention to the mathematical foundations of his gear-train algorithm. (In a longer essay, which I have still not been able to lay hands on, Brocot claims to provide more theoretical background.) There is no sign that either man knew of the other's work. After the fact, however, connections between them are easy to see; they are doing the same thing but describing it differently.

There is even a connection between Brocot's algorithm and the Fibonacci series, where this whole quest began. To see the relation, just try using Brocot's method to find ratios approximating the constant known as φ, or the golden ratio, an irrational number with a value of approximately 1.618. The series of approximants begins $\frac{1}{1}$, $\frac{2}{1}$, $\frac{3}{2}$, $\frac{5}{3}$, $\frac{8}{5}$, $\frac{13}{8}$, ... Hidden within these ratios is the complete sequence of Fibonacci numbers.

Working through examples of Brocot's process by hand, and leafing through the pages of the printed Brocot table, leaves me feeling wistful and uneasy. The ingenuity and diligence on exhibit here are certainly admirable, and yet from a modern point of view they are also tinged with a horrifying futility. I am reminded of those prodigies who spent years of their lives calculating digits of the decimal expansion of π—a task that is now a mere warm-up exercise for computer software. I cannot help wondering which of my own labors will appear equally quaint and pathetic to some future reader who ransacks libraries for old volumes like this one.

The fact is, the design of simple gear trains is no longer a computationally interesting problem, because computation itself has overwhelmed it. With so much calculating power at your fingertips, it's hardly worth the bother of being clever.

You can solve gearing problems by brute force, using methods that would have been unthinkable for Camus or Brocot, or even for Merritt, who was writing only sixty years ago. If you need to approximate a ratio, just have the computer try all pairs of gears with no more than one hundred teeth. There are only 10,000 combinations; you can churn them out in an instant. For a two-stage compound train, running through the 100 million possibilities is a labor of minutes. The whirling gears of progress have put the gear makers out of work.

AFTERTHOUGHTS

When I set out to write about the mathematics of gear ratios, I thought the subject was mainly of historical interest. After the essay was published, however, a number of contemporary clockmakers stepped forward and taught me otherwise. Although many horology enthusiasts are primarily collectors of older mechanical timepieces, there are also some who design and build new gearing. They face the same kinds of problems that Brocot did.

One clockmaker (and astronomer) disputed my glib assertion that the search for optimal gear combinations is no longer a serious computational challenge. John G. Kirk of Santa Barbara, California, described the kind of problem he comes up against in practice: finding all combinations of six gears in a compound train, with each gear allowed to have any number of teeth between 15 and 256. There are more than 200 trillion such combinations, and I concede that a brute-force enumeration of them is totally impractical. (Combining some common sense with brute force does produce a solution.)

There is more to say on the history of the Stern-Brocot tree. Several readers mentioned that this arrangement of numbers looked familiar, but they knew it by a different name: the Farey sequence. John Farey (1766–1826) was a British geologist who apparently had just one publication on a mathematical topic, a letter published in 1816 in *The Philosophical Magazine and Journal*, where he described "a curious property of vulgar fractions." The property is essentially a definition of the mediant. This was more than forty years before either Stern or Brocot weighed in on the subject. But Farey was not the originator of the idea either. His letter refers to tables of fractions compiled by Henry Goodwyn, who evidently also knew of the mediant property. Earlier still, in 1802, someone named Haros wrote on the same theme in a French journal. (I've been unable to learn anything at all about Haros—not even his first name. Citations of his 1802 article often list him as "C. Haros," but the actual byline reads "le C.^{en} Haros"—that is, "Citizen Haros," a common mode of address in that revolutionary era.)

If Haros and Farey got there first, why speak of the Stern-Brocot tree? In my view, the two ideas are not quite the same, although the distinctions are subtle. For one thing, Haros and Farey considered only the fractions in the interval between 0 and 1, whereas Stern and Brocot cover all positive rational numbers. More important, Haros and Farey thought of their method as producing a linear sequence of fractions arranged in order of increasing magnitude. Stern and Brocot described not a sequence but a tree—a mechanism for generating all possible rational numbers.

Another historical detail was brought to my attention by Donald E. Knuth, the distinguished computer scientist at Stanford University. Knuth pointed out that Stern himself did not claim credit for the idea behind the tree of rationals; he attributed it to Gotthold Eisenstein, a younger friend and colleague. In Eisenstein's collected works, Knuth found an 1850 letter to Stern that set forth the basic idea of the tree—although it was Stern who supplied the mathematical proofs.

Enclosed with Knuth's note to me was the trophy that I reproduce on the previous page, and which I mention with a mixture of pride and chagrin. For many years Knuth has offered such bounties to anyone who reports an error in his work. Strictly speaking, I never did *report* an error; I merely mentioned in passing that an inexact citation had sent me to the wrong volume of the *Revue chronométrique*. That oblique comment was all it took. Given my own lapses in scholarly diligence—including some that I document in this essay—I am slightly embarrassed to accept Knuth's reward. Then again, no one would ever cash such a check.

Horacio Caruso, whose own scholarly diligence put me on the trail of Stern and Brocot in the first place, died in 2003. He was a dear friend, though we never met. His protégé Sebastián Marotta is now at Boston University.

CHAPTER 8

The Easiest Hard Problem

One of the cherished customs of childhood is choosing up sides for a ball game. Where I grew up, we did it this way: The two chief bullies of the neighborhood would appoint themselves captains of the opposing teams, and then they would take turns picking other players. On each round, a captain would choose the most capable (or, toward the end, the least inept) player from the pool of remaining candidates, until everyone present had been assigned to one side or the other. The aim of this ritual was to produce two evenly matched teams and, along the way, to remind each of us of our precise ranking in the neighborhood pecking order. It usually worked.

None of us in those days—not the hopefuls waiting for our names to be called, and certainly not the two thick-necked team leaders—recognized that our scheme for choosing sides implements a greedy heuristic for the balanced number-partitioning problem. And we had no idea that this problem is NP-complete—that finding the optimum team rosters is certifiably hard. We just wanted to get on with the game.

And therein lies a paradox: If computer scientists find the partitioning problem so intractable, how come children the world over solve it every day? Are the kids *that* much smarter? Quite possibly they are. On the other hand, the success of playground algorithms for partitioning might be a clue that the task is not always as hard as that forbidding term "NP-complete" tends to suggest. As a matter of fact, finding a hard instance of this famously hard problem can be a hard problem—unless you know where to look. Some recent results, which make use of tools borrowed from physics and mathematics as well as computer science, show exactly where the hard problems hide.

The organizers of sandlot ball games are not the only ones with an interest in efficient partitioning. Closely related problems arise in many other resource-allocation tasks. For example, scheduling a series of computing jobs on a machine with dual processors is a partitioning problem: sorting the jobs into two sets with equal running time will balance the load on the processors. Another example is apportioning the miscellaneous assets of an estate between two heirs.

So What's the Problem?

Here is a slightly more formal statement of the partitioning problem. You are given a set of n positive integers, and you are asked to separate them into two subsets; you may put as many or as few numbers as you please in each of the subsets, but you must make the sums of the subsets as nearly equal as possible. Ideally, the two sums would be exactly the same, but this is feasible only if the sum of the entire set is even; in the event of an odd total, the best you can possibly do is to choose two subsets that differ by 1. Accordingly, a perfect partition is defined as any arrangement for which the "discrepancy"—the absolute value of the subset difference—is no greater than 1.

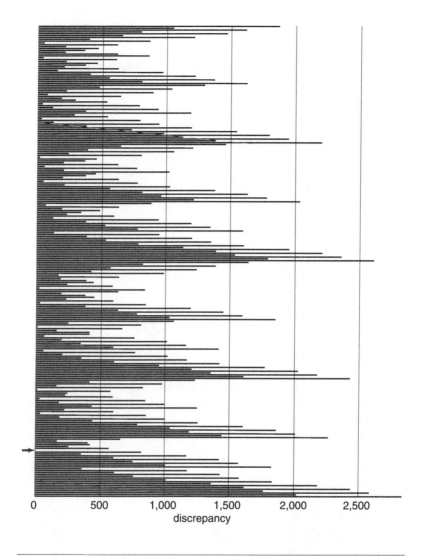

0   500   1,000   1,500   2,000   2,500

discrepancy

One perfect partition hides in a dense thicket of very imperfect ones. Partitions are ways of dividing a set of numbers into two subsets; a partition is perfect if the subsets have the same sum. Here the bars represent the discrepancy—the absolute value of the subset difference—of the 256 ways of partitioning a certain set of nine integers. The set chosen for the example is {484 114 205 288 506 503 201 127 410}; the lone perfect partition, marked by an arrow in the left margin, divides the numbers into the subsets {410 503 506} and {127 201 288 205 114 484}, each of which adds up to 1,419.

Try a small example. Here are ten numbers—enough for two basketball teams—selected at random from the range between 1 and 10:

$$\{2\ 10\ 3\ 8\ 5\ 7\ 9\ 5\ 3\ 2\}$$

Can you find a perfect partition? In this instance it so happens there are twenty-three ways to divvy up the numbers into two groups with exactly equal sums (or forty-six ways if you count mirror images as distinct partitions). Almost any reasonable method will converge on one of these perfect solutions. This is the answer I stumbled onto first:

$$\{2\ 5\ 3\ 10\ 7\}\quad\{2\ 5\ 3\ 9\ 8\}$$

Both subsets sum to 27.

This example is in no way unusual. As a matter of fact, among all sets of ten integers between 1 and 10, more than 99

set to be partitioned	left	right	discrepancy
19 17 13 9 6			
17 13 9 6	19		19
13 9 6	19	17	2
9 6	19	17 13	11
6	19 9	17 13	2
	19 9 6	17 13	4

The greedy algorithm is one of the simplest approximate methods of partitioning. The rule is always to take the largest number remaining to be assigned and put it in the subset with the smaller sum. Here gray balls indicate the number being moved at each stage.

percent have at least one perfect partition. (To be precise, of the 10 billion such sets, 9,989,770,790 can be perfectly partitioned. I know because I counted them—and it wasn't easy.)

Maybe larger sets are more challenging? With a list of a thousand numbers between 1 and 10, working the problem by pencil-and-paper methods gets tedious, but a simple computer program makes quick work of it. A variation on the two-bullies algorithm does just fine. First sort the list of numbers according to magnitude, then go through them in descending order, assigning each number to whichever subset currently has the smaller sum. This is called a greedy algorithm, because it takes the largest numbers first.

The greedy algorithm almost always finds a perfect partition for a list of a thousand random numbers no greater than 10. Indeed, the procedure works equally well on a set of ten thousand or a hundred thousand or a million numbers in the same range. The explanation of this success is not that the algorithm is a marvel of ingenuity. Lots of other methods do as well or better.

A particularly clever algorithm was described in 1982 by Narendra Karmarkar and Richard M. Karp, who were then at the University of California, Berkeley. It is a "differencing" method: at each stage you choose two numbers from the set to be partitioned and replace them by the absolute value of their difference. This operation is equivalent to deciding that the two selected integers will go into different subsets, without making an immediate commitment about which numbers go where. The process continues until only one number remains in the list; this final value is the discrepancy of the partition. You can reconstruct the partition itself by working backward through the series of decisions. In the search for perfect partitions, the Karmarkar-Karp procedure succeeds even more often than the greedy algorithm.

set to be partitioned	difference	left	right	discrepancy
19 17 13 9 6	2	13 19	17 9 6	0
13 9 6 2	4	13 2	9 6	0
6 4 2	2	4 2	6	0
2 2	0	2	2	0
0			0	0

The Karmarkar-Karp algorithm operates in two phases. First, reading down the left-hand side of the diagram, pairs of numbers are replaced by their difference, effectively deciding they will go into different subsets. In the second phase, reading up the right-hand side, the partition is constructed from the sequence of differencing decisions. Here the 0 at the bottom of the table derives from the difference of two 2s, which can therefore be inserted, one in each subset. One of the 2s arose as the difference between a 6 and a 4, so those numbers can also be written down, and so on.

At this point you may be ready to dismiss partitioning as just a wimpy problem, unworthy of the designation NP-complete. But try one more example. Here is another list of ten random numbers, chosen not from the range 1 to 10 but rather from the range between 1 and 2^{10}, or 1,024:

$$\{771 \quad 121 \quad 281 \quad 854 \quad 885 \quad 734 \quad 486 \quad 1{,}003 \quad 83 \quad 62\}$$

This set *can* be partitioned perfectly, but there is just one way to do it, and finding that unique solution takes persistence. The greedy algorithm does not succeed; it gets stuck on a partition with a discrepancy of 32. Karmarkar-Karp does slightly better, reducing the discrepancy to 26. But the only sure way to find the one perfect partition is to check all possible partitions, and there are 1,024 of them.

If this challenge is still not daunting enough, try partitioning one hundred numbers ranging up to 2^{100}, or a thousand numbers up to $2^{1,000}$. Unless you get very lucky, deciding whether such a set has a perfect partition will keep you busy for quite a few lifetimes.

Where the Hard Problems Are

To make sense of what's going on here, we need to be clear about what it means for a problem to be hard. Computer science has reached a rough consensus on this issue: easy problems can be solved in "polynomial time," whereas hard problems require "exponential time." If you have a problem of size x and you know an algorithm that can solve it in x steps or x^2 steps or even x^{50} steps, then the problem is officially easy; all of these expressions are polynomials in x. But if your best algorithm needs 2^x steps or x^x steps, you're in trouble. Such exponential functions (where x appears as an exponent) grow faster than any polynomial for large enough values of x.

The easy, polynomial-time problems are said to lie in class P; hard problems are all those not in P. The notorious class NP consists of some rather special hard problems. As far as anyone knows, solving these problems requires exponential time. On the other hand, if you are given a proposed solution, you can check its correctness in polynomial time. (NP stands for "non-deterministic polynomial"—although admittedly that's not much help in understanding the concept.)

Where does number partitioning fit into this taxonomy? Both the greedy algorithm and Karmarkar-Karp have polynomial running time; they can partition a set of n numbers in fewer than n^2 steps. For purposes of classification, however, these algorithms simply don't count, because they're not guaranteed to find the right answer. The hierarchy of problem difficulty is based on a worst-case analysis, which disqualifies an algorithm if it fails on even one problem instance. The only known method that passes the worst-case test is the brute-force approach of examining every possible partition. But this is an exponential-time algorithm: each integer in the set can be assigned to either of the two subsets, so that there are 2^n partitions to be considered.

Partitioning is a classic NP problem. If someone hands you a list of n numbers and asks, "Does this set have a perfect partition?" you can always find the answer by exhaustive search, but this can take an exponential amount of time. If you are given a proposed perfect partition, however, you can easily verify its correctness in polynomial time. All you need to do is add up the two subsets and compare the sums, which takes time proportional to n.

Indeed, partitioning is not just an ordinary member of the class NP; it is one of the elite NP problems designated NP-complete. What this means is that if someone discovered a polynomial-time algorithm for partitioning, it could be adapted to solve all NP problems in polynomial time. Each NP-complete problem is a skeleton key to the entire class NP.

Given these sterling credentials as a hard problem, it's all the more perplexing that partitioning often yields so readily to unsophisticated methods such as the greedy algorithm. Does this problem have teeth, or is it just a sheep in wolf's clothing?

Boiling and Freezing Numbers

An answer to this question has emerged in recent years from a campaign of research that spans at least three disciplines—physics, mathematics, and computer science. It turns out that the spectrum of partitioning problems has both hard and easy regions, with a sharp boundary between them. On crossing that frontier, the problem undergoes a phase transition, analogous to the boiling or freezing of water.

In general, problems get harder as they get bigger. If you're adding up a column of numbers, you can expect to do more work if there are more numbers or if the individual numbers have more digits. The size of a partitioning problem can be measured in the same way: the size is equal to n, the number of integers in the set, multiplied by m, the number of digits in a typical integer. The custom is to express m in terms of binary digits, or bits. Thus a partitioning problem with 100 integers of 10 bits each has a size of 1,000 bits.

Other things being equal, partitioning problems do get harder as their size increases. Surprisingly, though, the product of n and m is not the best predictor of difficulty in number partitioning. Instead, it's the ratio $^m/n$.

Some simple reasoning about extreme cases takes the mystery out of this assertion. Suppose the ratio of m to n is very small, say with $m = 1$ and $n = 1,000$. The task, then, is to partition a set of 1,000 numbers, each of which can be represented by a single bit. This is trivially easy: a one-bit integer greater than 0 can have only one possible value, namely 1, and so the input to the problem is a list of a thousand 1s. Finding a perfect partition is just a matter of counting. At the opposite extreme, consider the case of $m = 1,000$ and $n = 2$. Here the separation into subsets is easy—how many ways can you partition a set of two items?—but the likelihood of a perfect partition is extremely low. It's the probability that two randomly selected 1,000-bit numbers will be equal.

Now it becomes clear why in my baby-boom neighborhood we so easily formed ourselves into well-matched teams. Among the dozen or more kids who would gather for a game, athletic abilities may have differed by a factor of 10, but surely not by a factor of 1,000. The parameter m is the base-2 logarithm of

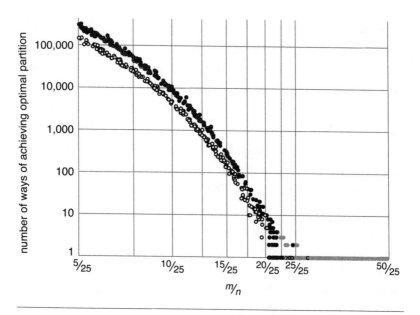

An abundance of solutions explains why some partitioning problems are embarrassingly easy. The crucial parameters are the size of the set, n, and the size of a typical member of the set, m, measured in bits. When m is less than n, most sets have many perfect partitions. When m is greater than n, the best partition is usually unique and is seldom perfect. Plotted here are the number of optimal partitions for sets with n = 25 and m ranging from 5 to 50. Imperfect partitions are gray; perfect partitions are black, with solid dots for odd-sum sets and open circles for even-sum sets. There is a phase transition near n = m.

this range of talents, and so it was no greater than 3 or 4. Thus $^m/_n$ was rather small, and we had many acceptable solutions to choose from.

Antimagnetic Numbers

The ratio $^m/_n$ divides the space of partitioning problems into two regions. Somewhere between them—between the fertile valley where solutions bloom everywhere and the stark desert where even one perfect partition is too much to expect—there must be a crossover region. There lies the phase transition.

The concept of a phase transition comes from physics, but it also has a long history of applications to mathematics, especially in the area known as combinatorics, where the main business is counting how many ways things can be combined or arranged. Forty years ago Paul Erdős and Alfréd Rényi described phase transitions in the growth of random graphs (collections of dots and connecting lines). Suppose you start with a set of isolated dots, and for each pair of dots you decide randomly but with a fixed probability whether or not to draw a line between them. Erdős and Rényi discovered that as the probability increases, there is a sudden transition. Below the threshold, most of the dots get connected only to a few neighbors; above the threshold, the network of connecting lines becomes dense enough that you can almost always find a path from any dot to any other. It's the suddenness of this change that suggests comparisons with a phase transition. The nature of the graph shifts abruptly at the threshold, just as the state of water changes dramatically when it goes from 99 degrees Celsius to 101 degrees.

By the 1980s, phase transitions had been observed in many combinatorial processes. The most thoroughly explored example is an NP-complete problem called satisfiability. A 1991 article by Peter Cheeseman, Bob Kanefsky, and William M. Taylor, titled "Where the Really Hard Problems Are," conjectured that all NP problems have a phase transition, and suggested that this is what distinguishes them from problems in P.

Meanwhile, in another paper with a provocative title ("The Use and Abuse of Statistical Mechanics in Computational Complexity"), Yaotian Fu of Washington University in St. Louis argued that number partitioning is an example of an NP problem without a phase transition. This assertion was disputed by Ian P. Gent (now of the University of St. Andrews in Scotland) and Toby Walsh (now of the University of New South Wales in Australia), who presented strong computational evidence for the existence of a phase transition in number partitioning. Their measurements suggested that the critical value of the $^m/n$ ratio, where easy problems give way to hard ones, is about 0.96.

Stephan Mertens of the Otto von Guericke Universität in Magdeburg, Germany, has given a thoroughgoing analysis of number partitioning from a physicist's point of view. His survey paper (which has been my primary source in writing this essay) appears in a special issue of *Theoretical Computer Science* devoted to phase transitions in combinatorial problems.

As a means of understanding the phase transition, Mertens sets up a correspondence between the number-partitioning problem and a model of a physical system. To see how this works, it helps to think of the partitioning process in a new context. Instead of unzipping a list of numbers into two separate lists, keep all the numbers in one place and multiply some of them by −1. The idea is to negate just the right selection of numbers so that the entire set sums to zero. Now comes the leap from mathematics to physics: the collection of positive and negative numbers is analogous to an array of atoms in a magnetic material, with the plus and minus signs representing atoms whose magnetic fields point in opposite directions. Physicists describe magnetic substances in terms of atomic "spins" that point either up or down; the two orientations correspond roughly to the more familiar north and south poles of a magnet. In a material known as a ferromagnet—this is what you use to stick notes on the refrigerator door—most of the spins are lined up in the same direction. The array of spins that solves the number-partitioning problem is that of an *antiferromagnet*, a more exotic material in which neighboring spins prefer to point in opposite directions. Specifically, the number-partitioning system resembles an infinite-range antiferromagnet, where every spin can feel the influence of every other spin, no matter how far apart they are.

The idea of studying partitioning by turning it into a problem about magnets may seem slightly perverse. It takes a simply stated problem in combinatorics and turns it into a messy and rather obscure system in the department of physics known as statistical mechanics. Why bother? The reason is that physics offers some powerful tools for predicting the behavior of such a system. In particular, the interplay of energy and entropy governs how the collection of spins can be expected to evolve toward a state of stable equilibrium. Physical systems usually tend to minimize energy and maximize entropy (which is a measure of disorder). In this case, the energy comes from the interaction between atomic spins (or between positive and negative numbers); because the system is an antiferromagnet, the energy is minimized when the spins point in opposite directions (or when the numbers sum to zero). The entropy depends on the number of ways of achieving the minimum-energy state. If there is just one way of arranging the spins so that all the ups and downs are balanced, then the entropy is zero. Likewise, a set of numbers with just one perfect partition also has zero entropy. When there are many equivalent ways of balancing the spins (or partitioning a set perfectly), the entropy is high.

The ratio m/n controls the state of this system. When m is much greater than n, the spins almost always have just one configuration of lowest energy. At the other pole, when m is much smaller than n, there are a multitude of zero-energy states, and the system can land in any one of them. Mertens showed that the transition between the two phases comes at $m/n = 1$, at least when n is very large and tending toward infinity. He derived corrections for finite n that may explain why the experiments of Gent and Walsh gave the slightly different transition point 0.96.

Finally, Mertens showed just how hard the partitioning problems become on the hard side of the phase transition. Searching for the best partition in this region is equivalent to searching a randomly ordered list of random numbers for the smallest element of the list. Only an exhaustive traverse of the entire list can guarantee an exact result. What's worse, there are no really good heuristic methods; no shortcuts are inherently superior to blind, random sampling.

Yet this is not to say that heuristics for partitioning are totally worthless. On the contrary, the phase-transition model helps explain how the Karmarkar-Karp algorithm works. The differencing operation reduces the range of the numbers and so diminishes $^m/n$, sliding the problem toward the easy phase. The differencing algorithm can't be counted on to find the very best partition, but it's an effective way of avoiding the worst ones.

Back to Mathematics

The work of Mertens has answered most of the major questions about number partitioning, and yet it's not quite the end of the story. The methods of statistical mechanics were developed as tools for describing systems made up of vast numbers of component parts, such as the atoms of a macroscopic specimen of matter. When applied to number partitioning, the methods are strictly valid only for very large sets, where m and n both go to infinity (while maintaining a fixed ratio). Those who need to solve practical partitioning problems are generally interested in somewhat smaller values of m and n.

Mertens's results have another limitation as well. In calculating the distribution of optimal solutions, he had to adopt a simplifying approximation at one crucial step, assuming that certain energies in the spin system are random. The true distribution of those energies is harder to deduce, but the task has been undertaken by Christian Borgs and Jennifer T. Chayes of Microsoft Research and Boris Pittel of Ohio State University. They have reclaimed the problem from the realm of physics and brought it back to mathematics. Their paper giving detailed proofs runs to nearly forty pages.

To their surprise, Borgs, Chayes, and Pittel discovered that Mertens's random-energy approximation actually yields exact results in the limiting case of infinite m and n. Under these conditions the phase transition is perfectly sharp. Anywhere below the critical ratio $^m/n = 1$, the probability of a perfect partition is 1; above this threshold the probability is 0. Borgs, Chayes, and Pittel also give a precise account of how the phase transition softens and broadens for smaller problems. When n is finite, the probability of a perfect partition varies smoothly between 0 and 1, with a "window" of finite width surrounding the critical ratio.

If my friends and I back on the ball field had known all this, would we have played better games? Probably not, but in retrospect I take satisfaction in the thought that our ritual for choosing teams was algorithmically well-founded.

AFTERTHOUGHTS

There's a story in one of the illustrations that accompany this essay—a story that I didn't get a chance to tell when the article was first published in *American Scientist*. I was still finishing some of the artwork on the night before the magazine was scheduled to go to press. The last of the illustrations (from which I reproduce a detail below) required a fair amount of computing; I had to run an exhaustive search for the best solutions of about a thousand partitioning problems. When I finally had the results in hand and started drawing graphs of the data, there was a puzzling feature that worried me. In these experiments, the size of the sets, n, was fixed at 25, and the size of the individual numbers, m, varied from 5 bits to 125 bits. As expected, the number of perfect partitions grew exponentially as the ratio m/n got smaller. The curve traced out by the data points was not perfectly smooth, but that was no surprise because I was working with individual random sets rather than

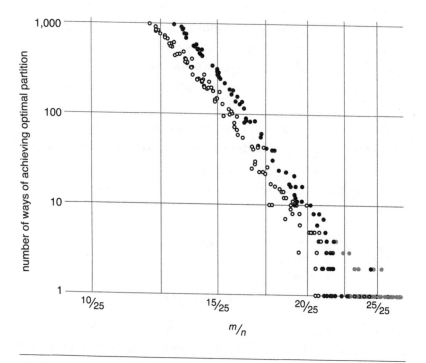

Computer experiments revealed that sets whose sum is an odd number (solid black dots) have about twice as many perfect partitions, on average, as similar sets whose sum is an even number (open circles).

averaging the results for large numbers of sets. I wasn't bothered by the jiggling and jittering of the curve. What perplexed
me was that the dots seemed to trace out two parallel curves
with a gap between them.

When I went back to the original data to investigate, I discovered an explanation that left me even more worried. According to my program, sets whose sum is an odd number have
about twice as many perfect partitions, on average, as sets with
an even sum. At a value of $^m/n$ where even sets have about a
hundred perfect solutions, odd sets have roughly two hundred.
On the logarithmic scale of the graph, this twofold difference
in abundance of solutions creates two parallel curves.

When I discovered this even-odd disparity, I was sure there
must be a bug in my program. I was probably missing half of
the solutions for even-sum sets, or else double counting the solutions for odd sets. I looked at the code but saw nothing
amiss. I ran more tests. I modified the program so that it not
only counted the perfect partitions but also printed them out;
checking these long lists did not reveal any obvious omissions
or duplications, and indeed there seemed to be about twice as
many solutions for the sets with odd sums. This was encouraging in one sense—maybe I hadn't goofed so badly—but I was
still in the dark about the underlying cause. Why should partitions of odd-sum sets be more numerous? I leafed through my
big heap of journal articles and preprints but found no explanation. I took a close look at graphs others had published, but all
of those authors averaged their data before plotting the points,
a process that would collapse the two parallel lines into one
along the mean value. It was the middle of the night. The press
deadline was looming.

This story has a happy ending because mathematicians and
physicists never sleep, and they answer e-mail at all hours. I
soon had help from Stephan Mertens, Christian Borgs, and

Jennifer Chayes, and further guidance from the expert proba-
bilists J. Laurie Snell of Dartmouth College and Charles Grin-
stead of Swarthmore College. The answer is that odd and even
sets really do behave differently. Suppose the sum S of a given
set of integers is an even number; then a perfect partition must
have a sum exactly equal to $S/2$. If S is odd, on the other hand,
subsets that add up to either $(S+1)/2$ or $(S-1)/2$ are considered perfect partitions. In effect, there are twice as many ways to construct a perfect partition of an odd-sum set. (If I had known
where to look and what to look for, I would have found this
fact already stated in a paper by Borgs, Chayes, and Pittel.)

After the article was published, Donald B. Aulenbach, an en-
vironmental engineering consultant on Long Island, pointed
out that my playground method of choosing up sides for a ball
game can be significantly improved by a very simple modifica-
tion. In the scheme I described, the two captains—call them A
and B—choose players in the order AB, AB, AB, AB, . . . Thus
A gets first pick in every round and can expect to wind up with
the stronger team. The obvious improvement is to alternate
priority: AB, BA, AB, BA, . . . An equivalent rule, as Aulenbach noted, is for A to make the first choice and thereafter let
each captain take two turns in a row. Computer experiments
clearly confirm the superiority of the ABBA rule over ABAB.

Finally, when I mentioned early in the essay that set-
partitioning algorithms might be useful in scheduling computer jobs, I was unaware that this application had actually
been explored long before most of the other work I mentioned.
Ronald L. Graham, now of the University of California, San
Diego, published an article titled "Bounds on Multiprocessing
Timing Anomalies" in 1969.

CHAPTER 9

Naming Names

Adam's only chore in the Garden of Eden was naming the
beasts and birds. The book of Genesis doesn't tell us whether
he found this task burdensome, but today the need to name
and number things has become a major nuisance. When you
try to choose a name for a new Internet domain or an e-mail
account, you're likely to discover that your first choice was
taken long ago. One Internet service tells me the name "brian"
is unavailable and suggests "brian13311" as an alternative. Perhaps I should think of this appellation in the same category as
Louis the Eighteenth or John the Twenty-third, but being
Brian the 13,311th seems to me a dubious distinction.

The challenge of inventing original names is particularly
acute when the name has to fit into a format that allows only a
finite number of possibilities. For example, the ticker symbols
that identify securities on the New York Stock Exchange can
be no more than three characters long, and only the twenty-six
letters of the English alphabet are allowed. The scheme imposes an upper limit of 18,278 symbols. If the day ever comes
that 18,279 companies want to be listed on the exchange, the

format will have to be expanded. And long before that absolute limit is reached, companies could have a hard time finding a symbol that bears any resemblance to the company name.

It's not just names that are scarce; we're even running out of numbers. A few years ago telephone numbers and area codes were in short supply, and so were the numbers that identify computers on the Internet. Those crises abated, but then attention turned to the Universal Product Code, the basis of the bar-code labels found on virtually everything sold in the United States and Canada. It seems the universe has more products than the UPC has universal product numbers. For that reason and others, the twelve-digit UPC standard is being supplanted by a thirteen-digit code, with provisions for adding a fourteenth digit. The "sunrise" date for this transition was January 1, 2005. The big moment came and went, and no one seemed to notice—not even supermarket checkout clerks. Two years later, everything in my kitchen cupboards still has twelve-digit labels, but I suppose I should be grateful to know there's room in the database for ten times as many flavors of Pop-Tarts.

Finishing Adam's Job

Names and numbers were causing trouble long before the Internet age. Biology had a naming crisis in the seventeenth and eighteenth centuries. The problem wasn't so much a shortage of names as a surfeit of them: plants and animals were known by many different names in different places. Then came the great reform of Carolus Linnaeus and his system of Latin binomials, identifying each organism by genus and species. The new scheme revolutionized taxonomy, not because there is any magic in Latin or in two-part names but because Linnaeus and his followers labored to preserve a strict one-to-one mapping between names and organisms. Official codes of biological nomenclature continue to enforce this rule—one name, one species—although rooting out synonyms and homonyms is a constant struggle.

Linnaeus himself named some six thousand species, and by now the number of living things in the biological literature is approaching two million. But there could be another ten million species—or, who knows, even a hundred million—yet to be cataloged. Might we run out of names before all the living things are described? If we were to insist that every binomial consist of two real Latin words—words known to the Romans—then perhaps there might be trouble ahead. But in practice Linnaean names only have to *look* like Latin, and the only limit on their proliferation is the ingenuity of the biologist. A dictionary of classical Latin will not help you understand the terms *Nerocila* and *Conilera*, which designate two genera of isopods; more helpful is knowing that the biologist who invented the terms was fond of someone named Caroline.

Among all the sciences, the one with the most remarkable system of nomenclature is organic chemistry. Names in most other realms are opaque labels that identify a concept or object but tell you little about it. For most of us, a Linnaean name such as *Upupa epops* doesn't even reveal whether the organism is animal or vegetable (this one's a bird). In contrast, the full name of an organic compound specifies the structure of the molecule in great detail. For example, the name "1,1-dichloro-2,2-difluoro-ethane" is a prescription for drawing a picture of a Freon molecule. The mapping from name to structural diagram is so direct that it can be done by a computer program. The reverse transformation, from diagram to name, is trickier; in other words, it's easier to make the molecule from the name than to make the name from the molecule.

Exhausting the supply of names for organic compounds is not something we need to worry about: by the very nature of the notational system, there is a name for every molecule. On the other hand, the names can get so long and intricate that only a computer can parse them.

Namespace

Although difficulties with names are nothing new, the nature of name giving changed with the introduction of computer technology. There is greater emphasis now on making names uniform and unique. Moreover, many names and identifying numbers must conform to a rigid format, with a specified number of letters or digits drawn from a fixed alphabet.

Place-names—and abbreviations for them—offer a good example of how names have changed. In the old days, a letter from overseas addressed to the "U.S." or the "U.S.A." or even the "EE.UU." would stand a chance of being delivered, but electronic mail for the corresponding geographic domain must have the exact designation "US"; no variation is tolerated (except that uppercase and lowercase are not distinguished). The list of acceptable country codes for Internet addresses is maintained by the Internet Assigned Numbers Authority. Each code consists of exactly two characters, drawn from an alphabet of twenty-six letters. Thus the number of available codes—the total namespace—is 26×26, or 676. The current IANA list has 247 entries, so the filling factor—the fraction of the space that's occupied—is 0.365. That leaves room for growth if a few more nations decide to deconstruct themselves the way Yugoslavia and the Soviet Union did. But not every new nation can get its first-choice code.

Consider the case of the Åland islands, which used to be identified by a Web site within the FI (Finland) domain as "an autonomous, demilitarized and unilingually Swedish province of Finland." The islands are sufficiently autonomous to have

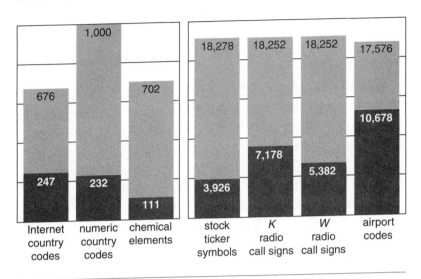

Constraints on the size and format of a name or numeric label create a
finite namespace, with room for only a fixed number of combinations.
The bar graphs show the total capacity of a few namespaces and the level
to which they are currently filled. (Left and right portions of the graph
have different scales.) The namespaces are two-letter country codes for
Internet domains, three-digit numeric country codes assigned by the
United Nations, the symbols of elements in the periodic table, ticker
symbols of stocks on the New York and American exchanges, call signs
of American radio stations beginning with K and W, and the three-
letter codes assigned to airports.

persuaded IANA to issue them a country code of their own—
but which code? Perhaps the first choice would have been AL,
but Albania already had that one. Or maybe AI, if Anguilla
hadn't claimed it. Why isn't Anguilla AN? Because that's the
code for the Netherlands Antilles (which might have been NA
if it weren't for Namibia). The preemption of AN also leads to
less-than-obvious assignments for Andorra, Angola, Antigua,
and even Antarctica. In the end, the Ålanders have wound up
with the code AX. If you try the old Finnish address, you'll be
redirected to www.aland.ax.

There is more to say about the difficulty of finding an un-used name as a namespace fills up. But first some more examples of finite namespaces:

Stock market ticker symbols. Ticker symbols began as telegraphers' informal shorthand, but today they are registered with the various exchanges. The New York Stock Exchange and the American Stock Exchange share a namespace; no symbol is allowed to have a different meaning in the two markets. If we ignore certain minutiae, the symbols consist of one, two, or three letters; the size of the namespace is $26^3 + 26^2 + 26 = 18{,}278$. The last time I checked, there were 3,926 active symbols, for a filling factor of about 0.22. Stocks traded on the NASDAQ market use four-letter symbols. There are fewer of these stocks (about 3,400) and a much larger namespace (456,976), so it should be considerably easier to find a symbol for a new company there. (A notable recent addition was Google, which chose the symbol GOOG.)

Telephone numbers. Telephone numbers in North America have ten decimal digits (including the area code), which suggests that the capacity of the namespace should be ten billion numbers. Under the rules prevailing through the 1980s, however, less than a tenth of those combinations were valid telephone numbers. The format of a phone number in those days was expressed as *NZX-NNN-XXXX*, where *N* represents any of the digits 2–9, *Z* the digits 0–1, and *X* any digit in the full range 0–9. That works out to about 819 million numbers. Even that quantity should be plenty; there are roughly 300 million telephones in use in the United States. Nevertheless, during the early 1990s the supply of numbers within many area codes came close to exhaustion. Although the crisis was often blamed on the proliferation of modems, fax machines, and cellular telephones, the real culprit was an inefficient scheme of allocation: if a telephone company had even one subscriber within a region, the company was assigned a block of 10,000 numbers. The main remedy was allocating numbers in smaller blocks, but along the way the grammatical rules defining a telephone number were relaxed, and the namespace expanded. Any combination of digits of the form *NXX-NXX-XXXX* is now a valid phone number, allowing some 6.4 billion possibilities. With careful conservation, the supply is expected to last until sometime in the 2030s.

Product codes. As in the telephone system, the shortage of Universal Product Codes is partly a matter of allocation policy. Although a UPC number has twelve digits (implying a maximum capacity of a trillion items), the first digit is a category code that in practice is almost always 0, and the final digit is a checksum used for detecting errors. Of the remaining ten digits, five identify the manufacturer and five the individual product. Because of this fixed structure, every manufacturer automatically gets a block of 100,000 item numbers, even though most companies need far fewer. The new thirteen-digit standard not only expands the total namespace by a factor of ten but also allows a more flexible division of resources. In particular, some companies are being given a longer manufacturer code and fewer item codes.

The new product-code standard isn't really new. The United States and Canada are merely acceding to another standard, called the European Article Number, that is already in use almost everywhere else in the world. (How quaint that the scheme known only in part of North America is the one labeled "Universal.") After the merger, the entire suite of product codes is supposed to be renamed the Global Trade Item Number. Most of the bar-code scanning devices at checkout counters have long been able to read the thirteen-digit EAN format, but in many cases the database in the back office could not handle the extra digit. While making the necessary conversions, retailers have been urged to allow space for a fourteen-digit version of the GTIN. Publishers and libraries are also being asked to renumber their world as the International Standard Book Number is expanded to thirteen digits and brought under the GTIN umbrella.

Social Security numbers. With nine-digit decimal numbers, there should be a billion possibilities. The Social Security Administration has excluded only a few of them ("No SSNs with an area number of '666' have been or will be assigned"), so that the actual size of the namespace appears to be 987,921,198. Some 420 million numbers have been issued since 1936, for a filling factor of about 0.4. The supply of numbers may well outlast the supply of funds to pay benefits.

Other countries have quite different systems for allocating numbers analogous to the U.S. Social Security number. For example, the Italian *codice fiscale* is not an arbitrary number assigned to a person but rather a string of alphanumeric symbols calculated from personal data such as name and date and place of birth. This scheme eliminates all concerns over running out of numbers, but it has another potential hazard: if the algorithm for calculating the *codici* is not chosen very carefully, two individuals may wind up with the same number.

Radio station call signs. Broadcast radio stations in the United States have call signs of either three or four letters, but the first letter is always either *K* or *W*. These rules create a namespace with room for 36,504 entries. I was surprised to discover how densely filled this space is. Combining the AM and FM bands (many stations broadcast on both), there are 12,560 call signs currently registered with the Federal Communications Commission, a filling factor of more than one-third.

Airport codes. When you check a bag at the airport, the luggage tag is marked with a three-letter code that indicates where, if all goes well, you'll eventually retrieve your belong-

ings. The codes are administered by the International Air
Transport Association. There's a code for every airport that
has airline service, not to mention a few bus and train stations.
Surprisingly, the IATA codes are the most densely packed of
all the naming schemes I have encountered. Out of 17,576 possible codes, 10,678 are taken, a filling factor of 0.6. This may be
why some of the codes are less than obvious (YYC for
Calgary?), although many such minor mysteries have historical
explanations. Chicago's O'Hare airport is ORD because it was
once called Orchard Field.

Making Hash of a Name

Suppose you've just built a new airport or radio station or
founded a sovereign nation, and you want to register an identifying code with the appropriate agency. What is the likelihood
that your first choice will be available? Or your second or third
choice? How do these probabilities change as the namespace
fills up?

If we can make the assumption that preferences for codes
are distributed randomly throughout the namespace, then the
question is easily answered. The probability that your first
choice is already taken is just the filling factor of the namespace. The probability that both your first choice and your second choice are taken is the square of the filling factor, and so
on. For example, if the namespace is two-thirds filled, then in
two-thirds of the cases a randomly chosen code will already be
present; four-ninths of the time, two randomly generated codes
will both be taken.

Searching at random for an unused name is related to the
process known in computer science as hashing. The idea of
hashing is to store data items for quick retrieval by scattering
them seemingly at random throughout a table in computer

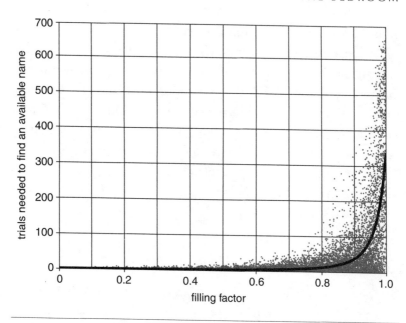

Simulation of the filling of a namespace suggests that finding a unique name becomes impractically difficult when the space is much more than half full. Starting with an empty namespace of 676 slots, the simulation adds names one at a time. First a name is generated at random; if the corresponding slot is empty, it is marked as filled. If the slot is already filled, the program tries the next slot in alphabetic sequence and continues in this way (wrapping around to the beginning of the space if necessary) until coming to an empty slot. For each name added, a dot is marked on the graph at the horizontal position corresponding to the filling factor at that moment and at the vertical position indicating the number of slots checked. The spray of gray dots superimposes forty repetitions of the process; the black line averages 100,000 trials.

memory. The arrangement isn't truly random; each item's position is set by a deterministic "hash function." Sometimes the hash function sends two data items to the same location; the collision must be resolved by putting one of the items elsewhere. This is analogous to requesting your favored name or code and finding that someone else has already claimed it.

The resemblance between name search and hashing is worth noting because the performance of various hashing algorithms has been carefully analyzed and documented. Much depends on the strategy for resolving collisions, or, in the context of name search, the policy for choosing an alternative when a desired name is not available. The illustration on the opposite page shows the results of a simulation of a name search equivalent to one of the simplest hashing methods. The rule here is to generate a first-choice name at random; if that choice is taken, try the next name in alphabetical order and continue until an opening is found. Naturally, the number of collisions increases as the namespace fills up, but the increase is not linear; the shape of the curve is concave upward. At a filling factor below about one-half, there is a reasonable chance you will get one of your first few choices. At higher filling factors, the average number of attempts before you find an available name rises steeply.

But there is a flaw in this analysis: the assumption that preferences for names are random is obviously bogus. People prefer names that appear to mean something or that have some trait that distinguishes them from random strings of symbols. In the stock market, the rare one-letter ticker symbols carry much prestige; radio call signs that spell a pronounceable word (WARM, KOOL) are in demand. It would be difficult to codify or quantify these biases, but as a simple way of estimating their effect, I tried looking at the first-order statistics of the code words in various data sets. The first-order statistics are simply the letter frequencies at each position within a word. (Higher-order statistics take into account correlations between the letters.)

My experiments compared the success of two players—one who chooses names utterly at random and another whose random choices are biased to match the statistics of the names already in the data set. In other words, the latter player tends to favor names that are like those already present. Not surprisingly, the random player has an easier time finding an available name. The magnitude of the effect can be quite large. In the case of IANA country codes, random choices succeed after an average of 1.6 probes, but finding a name with letter frequencies similar to the existing population takes 2.5 trials on aver-

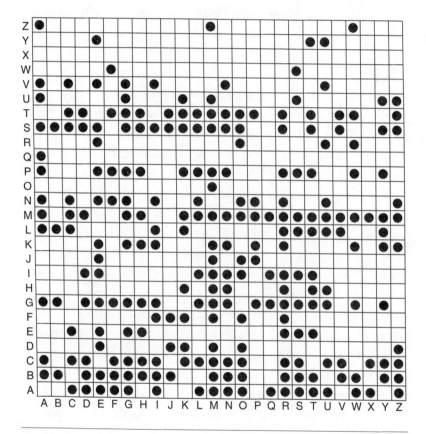

Names are distributed nonrandomly in real namespaces. The tableau has a dot for each two-letter country code in the list maintained by the Internet Assigned Numbers Authority. The table is read from bottom to top and left to right; thus AZ (Azerbaijan) is at the lower right and ZA (South Africa) at the upper left.

age. For IATA airport codes, the statistical bias raises the average number of attempts from 2.5 to 3.9. These results suggest that some namespaces may become impractically full much sooner than would be expected from an analysis based on hashing algorithms.

The experiment itself has a curious bias. Using an existing data set to infer people's preferences neglects the fact that many of the code words may not have been anyone's first choice; they may have been selected merely because the real first choice was already taken. Furthermore, the statistical bias varies with the filling factor. If there are only a few names in the data set, the letter frequencies will be strongly biased. Indeed, some letters may not appear at all in any name, and so the algorithm would assign those letters a probability of zero. At the opposite end of the spectrum, variations in letter frequencies inevitably diminish as the namespace fills up. Once almost all possible words are taken, all letters must have nearly the same frequency.

Horse Sense

As namespaces get larger, analyses based on random character strings become less illuminating. A case in point is the naming of thoroughbred horses. Under rules enforced by the Jockey Club, a horse's name can have from two to eighteen characters, drawn from an alphabet consisting of the usual twenty-six letters plus the space, the period, and the apostrophe. This is an enormous namespace, with room for more than 2×10^{26} entries. At any one time there are about 450,000 names assigned to active or recently retired horses. Most of these names will eventually become available for reuse, and so the pool of active names stays at roughly constant size. (Only the names of very famous steeds are permanently withdrawn; there will never be another Kelso or Secretariat.)

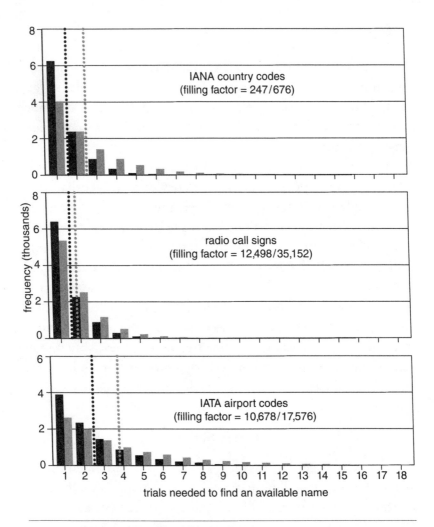

*Statistical bias within a namespace makes the search for an unclaimed
name even harder. Each of these graphs records the results of 10,000
independent attempts to add a single new name to an existing
namespace. The height of the bars indicates the number of times the
attempt succeeded on the first try, the second, the third, and so on. Black
bars are for names generated at random, gray bars for names with the
same first-order statistics as the names already in the data set. The
dotted black and gray lines give the average number of attempts needed
to find an open slot. In the case of airport codes, for example, it took
about 2.5 trials on average to find an unused random code but 3.9 trials
with codes that reflect the biased letter frequencies.*

With just 450,000 of 2×10^{26} slots occupied, the filling factor of this namespace might as well be zero. Generating strings of characters at random, you would have to try 10^{21} of them before you would have much chance of stumbling on a name in use. And yet real-world experience gives a very different impression. Of all the names submitted by horse breeders, the fraction rejected is not 1 in 10^{21} but close to 1 in 4. According to a spokesperson for the Jockey Club, the most common reason for rejection is that the proposed name is too close to an existing one. In this context names can clash even if they are not spelled identically—mere phonetic similarity is enough to bar a name. But even allowing for this broader criterion of uniqueness, the thoroughbred namespace is not nearly as empty as it would seem from a naive counting of character strings. A true estimate of the filling factor should probably be based not on the combinatorics of random letters but on combinations of words or some other higher linguistic unit.

The same is surely true for Internet domain names, such as www.bit-player.org or del.icio.us. Each component of a domain name—each part between dots—can have up to sixty-three characters, and the acceptable characters include letters and numbers as well as the hyphen. The size of the namespace is nearly 10^{100}; we won't use them all up anytime soon. But meaningful, pithy, clever domain names—that's another matter.

Even outside the confines of finite namespaces, the sheer onomastic challenge of modern life sometimes gets to be a burden. Where's Adam when we need him? Years ago, I could save a clipping from the newspaper without any need to name it. Now, for every document I create or download, I must enact a little ceremony of naming: I dub thee "FILE-037.TXT." The workload has gotten serious enough that consultants make a living out of nothing more than dreaming up names. (One firm named itself A Hundred Monkees—well named!)

When my daughter was a voluble three-year-old, she would greet passersby with the enthusiastic salute: "Hi! My name is named Amy. What is your name named?" A dizzying recursion yawns before us. Once we start naming names, and then the names of names of names, where do we ever stop?

AFTERTHOUGHTS

With the expansion of the Universal Product Code behind us, the next generation is already on the horizon. Under the new Electronic Product Code, items on store shelves will be identified not by printed bar codes but by RFID (radio-frequency identification) tags. With this technology, the tag doesn't have to be visible on the outside of the package; a scanner sends a triggering radio signal, and the tag responds with a radio-frequency message that includes the product code.

If RFID tags eventually catch on, they will bring a further dramatic expansion of the product namespace. Whereas the present standard has room for fourteen decimal digits, the Electronic Product Code specifies ninety-six binary bits, equivalent to between twenty-seven and twenty-eight decimal digits. That's a lot of numbers—but a lot will be needed, because the plan is to use them differently and more liberally. Existing product codes identify *categories* of objects: a certain UPC number specifies a fifteen-ounce box of Cheerios, and so every fifteen-ounce box of Cheerios carries this same number. The EPC will identify individual physical objects: each box of Cheerios will have its own serial number (or should I say cereal number?). The next step, presumably, will be to number the individual Cheerios.

Several variants of the EPC have been proposed. The one most closely related to the UPC system allocates thirty-eight bits to the serial number, which allows for counting up to 274,877,906,943 items.

In discussing chemical names, I mentioned that it's easier to deduce the structure of a molecule from a name than it is to go the other way, deriving the name from a diagram of the structure. Eugene Garfield let me know that although the latter task may indeed be difficult, there is computer software that succeeds in doing it. For example, the International Union of Pure and Applied Chemistry supplies software that converts a structural description into a label called an InChI, or international chemical identifier. The hydrocarbon known commonly as naphthalene has the InChI name 1/C10H8/c1-2-6-10-8-4-3-7-9(10)5-1/h1-8H. Within this long sequence of symbols the segment "C10H8" corresponds to the familiar compositional formula $C_{10}H_8$, indicating a molecule with 10 carbon atoms and 8 hydrogen atoms. The rest of the name encodes information about how the atoms are connected by chemical bonds.

Roy E. Plotnick pointed out that racehorses are not the only performers whose names are subject to approval by a sanctioning body. Professional actors face a similar prohibition on duplicated names, enforced by Actors' Equity and the Screen Actors Guild. If your birth certificate reads Cary Grant, you'll have to choose a new stage name when you take to the boards. (As far as I know, Archibald Leach is still available.)

The scarcity of meaningful names could have dramatic consequences in the case of stock ticker symbols, according to experiments by Adam L. Alter and Daniel M. Oppenheimer of Princeton University. They looked at the results of investing $1,000 in stocks with pronounceable symbols versus an equal investment in stocks with less-intelligible ticker codes. The one-day profit for the pronounceable stocks was $85 greater.

Many other namespaces could have been mentioned in this essay, and here I would like to call attention to three more. First are automobile license plates, including both standard-issue ones and vanity plates. Certain adjustments are needed in calculating the size of this namespace. In many cases, the letter I, the numeral 1, and even punctuation marks such as ! and : are considered identical on license plates, since a viewer might not be able to distinguish between them quickly. Other confusing pairs such as O and 0 may also be consolidated.

Second, astronomy has a serious naming problem. Stars and galaxies are so numerous there's really no hope of issuing meaningful or descriptive names; all but the most prominent objects get mere catalog numbers. Closer to the home planet, however, there is still an effort to humanize the sky. For example, the seventeenth-century Italian astronomer Giovanni Riccioli began the custom of naming craters on the moon after noted astronomers. (Crater Riccioli is a conspicuous one.) The trouble is, there are far more lunar craters than there are astronomers, and so the category had to be expanded. The International Astronomical Union now allows "famous deceased scientists, scholars, artists, and explorers."

Finally, my prize for the most wasteful use of a namespace goes to the shipping industry, which needs to keep track of all those freight containers you see stacked like LEGO bricks in ports and rail yards. Each of these boxes is stenciled with a four-letter code that identifies its owner, but the final letter is always U, and thus the code might as well be just three letters long.

CHAPTER 10

Third Base

People count by 10s and machines count by 2s—that pretty much sums up the way we do arithmetic on this planet. But there are countless other ways to count. Here I want to offer three cheers for base 3, the ternary system. The numerals in this sequence—beginning 0, 1, 2, 10, 11, 12, 20, 21, 22, 100, 101, 102—are not as widely known or widely used as their decimal and binary cousins, but they have charms all their own. They are the Goldilocks choice among numbering systems: when base 2 is too small and base 10 is too big, base 3 is just right.

How to Count

Most of the time, we don't distinguish very clearly between numbers and numerals—between the abstract concept of a quantity or magnitude and the symbols that represent the concept. The decimal numeral 19, the binary numeral 10011, and the Roman numeral XIX all refer to the same number. So does the even more primitive unary representation ||||| ||||| ||||| ||||. What's remarkable about this varied spectrum of counting systems is that they're all equivalent mathematically. That is, you can choose to do arithmetic with any of these numerals, and if you follow the rules correctly, you'll get the same answer—the same *number*.

The most important numerals are all constructed according to a place-value system. In decimal notation, the numeral 19 is shorthand for this expression:

$$(1 \times 10^1) + (9 \times 10^0),$$

or, as you might recite in a primary-school classroom, "one ten and nine ones." Likewise, the binary numeral 10011 is understood to mean:

$$(1 \times 2^4) + (0 \times 2^3) + (0 \times 2^2) + (1 \times 2^1) + (1 \times 2^0),$$

which adds up to the same value. The ternary version of the same number is written 201, which expands as follows:

$$(2 \times 3^2) + (0 \times 3^1) + (1 \times 3^0).$$

In this case we have two 9s, no 3s, and one 1.

The general formula for a numeral in any place-value notation goes something like this:

$$\ldots d_3 r^3 + d_2 r^2 + d_1 r^1 + d_0 r^0 \ldots$$

Here r is the base, or radix, and the coefficients d_i are the digits of the number. Usually, r is a positive integer and the digits are integers in the range from 0 to $r-1$, but neither of these restrictions is strictly necessary. (You can build perfectly good numbers on a negative or an irrational base, and below we'll meet numbers with negative digits.)

To say that all numerals represent the same numbers is not to say that all numeric representations are equally convenient for all purposes. Base 10 is famously well suited to those of us who count on our fingers. Base 2 dominates computing technology because binary devices are simple and reliable, with just two stable states—on or off, full or empty. Computer circuitry also exploits a coincidence between binary arithmetic and binary logic: the same signal can represent either a numeric value (1 or 0) or a logical value (true or false).

Cheaper by the Threesome

The cultural preference for base 10 and the engineering advantages of base 2 have nothing to do with any intrinsic mathematical properties of the decimal and binary numbering systems. Base 3, on the other hand, *does* have a genuine mathematical distinction in its favor. By one plausible measure, it is the most efficient of all integer bases; it offers the most economical way of representing numbers.

How do you measure the "cost" of a numeric representation? In answering this question, I find it helpful to think in terms of a mechanical counting device, such as an old-fashioned automobile odometer. The odometer might count from 000000 to 999999, for a total of a million distinct configurations. An odometer of this kind has six wheels, and each wheel has ten decimal digits engraved on its circumference. A reasonable measure of the counter's efficiency is the product of these two numbers: the width w (the number of wheels) multiplied by the radix r (the number of symbols per wheel). In this case the product wr is equal to 6×10, or 60. Note that the *capacity* of the counter—the number of things it can represent—is r^w, or 10^6.

Now suppose you buy a car with a binary odometer, where each indicator wheel bears only the two symbols 0 and 1. If this

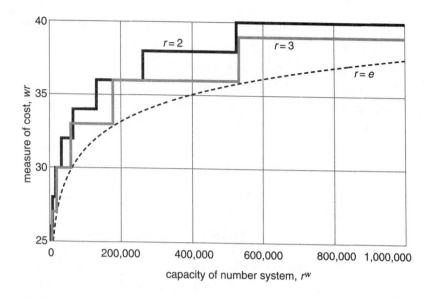

Base 3, the ternary system, generally yields the most efficient or economical representation of numbers. A numbering system is defined by the width w (the maximum number of digits in a number) and the radix r (the number of values each digit can assume). The capacity of the numbering system is r^w, and the cost of the representation is defined as the product wr. Among all integer values of r, setting r = 3 minimizes the cost for most values of r^w—indeed, for infinitely many of them. For 8,487 exceptions, r = 2 is the favored radix. If the radix is allowed to assume noninteger values, then r = e is the best choice. (The irrational number e is approximately 2.718.)

device is to record as many miles as the decimal version, it will need more wheels. Specifically, for a capacity of a million miles, the binary odometer needs twenty wheels, since 2^{20} is just over a million. Hence the cost of the binary representation is 20 × 2, or 40—substantially less than in the decimal case. A ternary odometer, where each wheel shows the symbols 0, 1, and 2, does even better. To reach the million-mile threshold, a ternary counter needs 13 digits, and so the cost function wr is equal to

13 × 3, or 39. It turns out that no other integer radix is more efficient than base 3 for this task.

The finding that 3 is the most efficient integer radix applies to the specific problem of representing the numbers from 0 through 999,999. Does the conclusion hold more generally? Yes and no. Suppose you want to build an odometer that registers from 0 up to 65,535 miles. A binary device can accomplish this with 16 digits (since 2^{16} is equal to 65,536). Therefore the cost wr for the binary counter is 16 × 2 = 32. A ternary counter needs 11 digits to reach the same total, and so its cost is 11 × 3 = 33. Here is an instance, then, where binary is more efficient than ternary, and so it cannot be true that base 3 is *always* the most efficient. But there are only finitely many such counterexamples. If you want to write all the numbers from 0 through N, binary notation is most efficient for 8,487 values of N, but ternary is superior for infinitely many other values.

Another way of comparing the cost and capacity of numbering systems is to treat the width w as a continuous variable rather than an integer, allowing for counters with a fractional number of digits. This strange-sounding notion has a simple rationale. The true capacity of a 20-digit binary counter is 2^{20}, or 1,048,576. When this device is used to count only as far as 999,999, some of the capacity goes to waste. The effective number of digits is not 20 but rather the base-2 logarithm of 1,000,000, which is approximately 19.93. Thus we might consider the binary counter a 19.93-digit device, and so the cost wr is reduced from 40 to 39.86. In a similar way, a 13-digit ternary counter has a capacity of 3^{13}, or 1,594,323. When the counter is restricted to a range of a million numbers, the effective width is the base-3 logarithm of 1,000,000, which is 12.58. The corresponding cost factor is 12.58 × 3 = 37.73. When the cost of a numbering system is expressed in this way, ternary is *always* the most efficient integer radix.

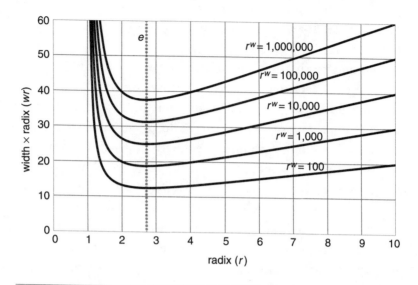

Comparisons of cost and capacity in numbering systems become more precise when the width w and the radix r are treated as continuous rather than integer variables. The most economical radix is e (about 2.718, marked by the dotted gray line). The uppermost curve gives the cost of representing a million numbers. For $r = 2$, the cost is 39.86; at $r = 3$, it is a little lower, 37.73; the minimum cost at $r = e$ is 37.55.

Admittedly, an odometer with 12.58 wheels is hard to envision. But if you can fathom that concept, you might as well contemplate the possibility that the radix, too, could be something other than an integer. If the aim is to minimize the product wr while keeping the quantity r^w constant, then the optimal value of r is e, the irrational number known as the base of the natural logarithms, with a numerical value of about 2.718. A million-mile odometer on which each wheel has readings from 0 up to e would need 13.82 wheels; multiplying e by 13.82 yields a cost function of 37.55—even lower than that of the ternary counter. In a sense, base 3 is the best of the integer bases because 3 is the integer closest to e.

Trit by Trit by Trit

This special property of base 3 attracted the notice of early computer designers. On the hypothesis that a computer's component count would be roughly proportional both to the width and to the radix of the numbers being processed, they suggested that wr might be a good predictor of hardware cost, and so ternary notation would make the most efficient use of hardware resources. The earliest published discussion of this idea I've been able to find appears in the 1950 book *High-Speed Computing Devices*, a survey of computer technologies compiled on behalf of the U.S. Navy by the staff of Engineering Research Associates.

At about the same time as the ERA survey, Herbert R. J. Grosch proposed a ternary architecture for the Whirlwind computer project at the Massachusetts Institute of Technology. Whirlwind evolved into the control system for a military radar network, which stood vigil over North American airspace through thirty years of the Cold War. Whirlwind was also the proving ground for several novel computer technologies—including magnetic core memory—but ternary arithmetic was not among the innovations tested; in spite of Grosch's recommendation, Whirlwind and its successors were binary machines.

As it happens, the first electronic ternary computer was built on the other side of the Iron Curtain. The machine was designed by Nikolai P. Brusentsov and his colleagues at Moscow State University and was named Setun, for a river that flows near the university campus. Some fifty of these machines were built between 1958 and 1965. Setun operated on numbers composed of 18 ternary digits, or trits, giving the machine a numerical range of 3^{18}, or 387,420,489. A binary computer would need 29 bits to reach this capacity; in terms of the cost function wr, the ternary design wins 54 to 58.

Unfortunately, the Setun machine did not realize the potential of base 3 to reduce component counts. Each trit was stored in a pair of magnetic cores, wired so that they had three stable states: both magnetized "up," both magnetized "down," or magnetized in opposite directions. A pair of cores could have held two binary bits, which amounts to more information than a single trit, and so the ternary advantage was squandered.

Along with ternary arithmetic, a computer built out of base-3 hardware can also exploit ternary logic. Consider the task of comparing two numbers. In a machine based on binary logic, comparison is often a two-stage process. First you ask, "Is x less than y?"; depending on the answer, you may then have to ask a second question, such as "Is x equal to y?" Ternary logic simplifies the process. A single comparison can yield any of three possible outcomes: "less," "equal," and "greater."

Ternary computers were a fad that faded, though not quickly. In the 1960s there were several more projects to build ternary logic gates and memory cells, and to assemble these units into larger components such as adders. In 1973 Gideon Frieder and his colleagues at the State University of New York at Buffalo designed a complete base-3 machine they called TERNAC, and they created a software emulator. Since then the idea of ternary computing has had occasional revivals, but you're not going to find a ternary laptop or minitower in stock at CompUSA.

Why did base 3 fail to catch on? One easy guess is that reliable three-state devices just didn't exist or were too hard to develop. And once binary technology became established, the tremendous investment in methods for fabricating binary chips would have overwhelmed any small theoretical advantage of other bases. Furthermore, it's only a hypothesis that such an advantage even exists. Everything hinges on the assumption that the product *wr* is a proper measure of hardware complexity, or in other words that the incremental cost of increasing the radix is the same as the incremental cost of increasing the number of digits.

But even if ternary circuits never find a home in computer hardware, the Goldilocks argument favoring base 3 may apply in other contexts. Suppose you are creating one of those dreadful telephone menu systems—press 1 to be inconvenienced, press 2 to be condescended to, and so forth. If there are many choices, what is the best way to organize them? Should you build a deep hierarchy with lots of little menus that each offer just a few options? Or is it better to flatten the structure into a few long menus? In this situation a reasonable goal is to minimize the number of options that the wretched caller must listen to before finally reaching his or her destination. The problem is analogous to that of representing an integer in positional notation: the number of items per menu corresponds to the radix *r*, and the number of menus is analogous to the width *w*. The average number of choices to be endured is minimized when there are three items per menu.

Turning to Ternary Dust

Although numbers are the same in all bases, some properties of numbers show through most clearly in certain representations. For example, you can see at a glance whether a binary number is even or odd: just look at the last digit. Ternary also distinguishes between even and odd, but the signal is subtler: a ternary numeral represents an even number if the numeral has an even number of 1s. (The reason is easy to see when you count powers of 3, which are invariably odd.)

More than twenty years ago, Paul Erdős and Ronald L. Graham published a conjecture about the ternary representation of powers of 2. They observed that 2^2 and 2^8 can be written in ternary without any 2s (the ternary numerals are 11 and 100111, respectively). But every other positive power of 2 seems to have at least one 2 in its ternary expansion; in other words, no other power of 2 is a simple sum of powers of 3. Ilan Vardi of the Institut des hautes études scientifiques has searched up to $2^{6,973,568,802}$ without finding a counterexample, but the conjecture remains open.

The digits of ternary numerals can also help illuminate a peculiar mathematical object called the Cantor set, or Cantor's dust. To construct this set, draw a line segment and erase the middle third; then turn to each of the resulting shorter segments and remove the middle third of those also, and continue in the same way. After infinitely many middle thirds have been erased, does anything remain? One way to answer this question is to label the points of the original line as ternary numbers between 0.0 and 0.222 . . . (Note that the repeating ternary fraction 0.222 . . . is exactly equal to 1.0, just as 0.999 . . . in decimal notation is also merely another way of writing 1.0.) Given this labeling, the first middle third to be erased consists of those points with coordinates between 0.1 and 0.122 . . . , or in other words all coordinates with a 1 in the first position after the radix point. Likewise, the second round of erasures eliminates all points with a 1 in the second position after the radix point. The pattern continues, and the limiting set consists of points that have no 1s anywhere in their ternary representation. In the end, almost all the points have been wiped out, and yet an infinity of points remain.

The Jewel in the Triple Crown

"Perhaps the prettiest number system of all," writes Donald E. Knuth in *The Art of Computer Programming*, "is the balanced ternary notation." As in ordinary ternary numbers, the digits of a balanced-ternary numeral are coefficients of powers of 3, but instead of coming from the set {0 1 2}, the digits are −1, 0, and 1. They are "balanced" because they are arranged symmetrically about 0. The negative digits are usually written with a vinculum, or overbar, instead of a prefixed minus sign, thus: $\bar{1}$.

As an example, the decimal number 19 is written $1\bar{1}01$ in balanced ternary, and this numeral is interpreted as follows:

$$(1\times3^3) - (1\times3^2) + (0\times3^1) + (1\times3^0),$$

or in other words $27 - 9 + 0 + 1$. Every number, both positive and negative, can be represented in this scheme, and each number has only one such representation. The balanced-ternary counting sequence begins 0, 1, $1\bar{1}$, 10, 11, $1\bar{1}\bar{1}$, $1\bar{1}0$, $1\bar{1}1$. Going in the opposite direction, the first few negative numbers are $\bar{1}$, $\bar{1}1$, $\bar{1}0$, $\bar{1}\bar{1}$, $\bar{1}11$, $\bar{1}10$, $\bar{1}1\bar{1}$. Negative values are easy to recognize because the leading trit is always negative.

The idea of balanced or signed number systems has quite a tangled history. Both the Setun machine and the Frieder emulator were based on balanced ternary, and so was Grosch's proposal for the Whirlwind project. In 1950 Claude E. Shannon published an account of symmetrical signed-digit systems, including ternary and other bases. But none of these twentieth-century inventors was the first. In 1840 Augustin Cauchy discussed signed-digit numbers in various bases, and Léon Lalanne immediately followed up with a discourse on the special virtues of balanced ternary. Twenty years earlier, John Leslie's remarkable *Philosophy of Arithmetic* had set forth methods of calculating in any base with either signed or unsigned digits. Leslie in turn was anticipated a century earlier by John Colson's brief essay on "negativo-affirmative arithmetick." Earlier still, Johannes Kepler used a balanced-ternary scheme modeled on Roman numerals. There is even a suggestion that signed-digit arithmetic was already implicit in the Hindu Vedas, which would make the idea very old indeed!

What makes balanced ternary so pretty? It is a notation in which everything seems easy. Positive and negative numbers are united in one system, without the bother of separate sign bits. Arithmetic is nearly as simple as it is with binary numbers; in particular, the multiplication table is trivial. Addition and subtraction are essentially the same operation: to subtract, just negate one number and then add. Negation itself is also effortless: change every $\bar{1}$ into a 1, and vice versa. Rounding is mere truncation: setting the least-significant trits to 0 automatically rounds to the closest power of 3.

The best-known application of balanced-ternary notation is in mathematical puzzles that have to do with weighing. Given a two-pan balance, you are asked to weigh a coin known to have some integral weight between 1 gram and 40 grams. How many measuring weights do you need? A hasty answer would be six weights of 1, 2, 4, 8, 16, and 32 grams. If the coin must go in one pan and all the measuring weights in the other, you can't do better than such a powers-of-2 solution. If the weights can go in either pan, however, there's a ternary trick that works with just four weights: 1, 3, 9, and 27 grams. For instance, a coin of 35 grams—$110\bar{1}$ in signed ternary—will balance on the scale when weights of 27 grams and 9 grams are placed in the pan opposite the coin and a weight of 1 gram lies in the same pan as the coin. Every coin of no more than 40 grams can be weighed in this way.

Martha Stewart's File Cabinet

Some weeks ago, rooting around in files of old clippings and correspondence, I made a discovery of astonishing obviousness and triviality. What I found had nothing to do with the content of the files; it was about their arrangement in the drawer.

Imagine a fastidious office worker—a Martha Stewart of filing—who insists that no file folder lurk in the shadow of another. The protruding tabs on the folders must be arranged so that adjacent folders always have tabs in different positions. Achieving this staggered arrangement is easy if you're setting up a whole new file, but it gets messy when folders are added or deleted at random.

A drawer filled with "half-cut" folders, which have just two tab positions, might initially alternate left-right-left-right. The pattern is spoiled, however, as soon as you insert a folder in the middle of the drawer. No matter which type of folder you choose and no matter where you put it (except at the very ends of the sequence), every such insertion generates a conflict. Removing a folder has the same effect. Translated into a binary numeral with left = 0 and right = 1, the pristine file is the alternating sequence . . . 0101010101 . . . An insertion or deletion creates either a 00 or a 11—a flaw much like a dislocation in a crystal. Although in principle the flaw could be repaired—either by introducing a second flaw of the opposite polarity or by flipping all the bits between the site of the flaw and the end of the sequence—even the most maniacally tidy record keeper is unlikely to adopt such practices in a real file drawer.

In my own files I use third-cut rather than half-cut folders; the tabs appear in three positions: left, middle, and right. Nevertheless, I had long thought—or rather I had assumed without bothering to think—that a similar analysis would apply, and that I couldn't be sure of avoiding conflicts between adjacent folders unless I was willing to shift files to new folders after every insertion. Then came my Epiphany of the File Cabinet a few weeks ago: suddenly I understood that going from half-cut to third-cut folders makes all the difference.

It's easy to see why; just interpret the drawerful of third-cut folders as a sequence of ternary digits. At any position in any such sequence, you can always insert a new digit that differs from both of its neighbors. Base 3 is the smallest base that has this property. Moreover, if you build up a ternary sequence by consistently inserting digits that avoid conflicts, then the choice of which symbol to insert at a given position is always a forced one; you never have to make an arbitrary selection between two or more legal possibilities. Thus, as a file drawer fills up, it is not only possible to maintain perfect Martha Stewart order; it's actually quite easy.

Deletions, regrettably, are more troublesome than insertions. There is no way to remove arbitrary elements from either a binary or a ternary sequence with a guarantee that two identical digits won't be brought together. (On the other hand, if you're fussy enough to fret about the positions of tabs on file folders, you probably never throw anything away anyhow.)

The protocol for avoiding conflicts between third-cut file folders is so obvious that I assume it must be known to file clerks everywhere. But in half a dozen textbooks on filing— admittedly a small sample of a surprisingly extensive literature—I found no clear statement of the principle.

Strangely enough, my trifling observation about arranging folders in file drawers leads to some mathematics of wider interest. Suppose you seek an arrangement of folders in which you not only avoid putting any two identical tabs next to each other but also avoid repeating any longer patterns. In other words, you want to rule out not only 00 and 11 but also 0101 and 021021. Sequences that have no adjacent repeated patterns of any length are said to be "square free," by analogy to numbers that have no duplicated prime factors.

In binary notation, the one-digit sequences 0 and 1 are obviously square free, and so are 01 and 10 (but not 00 or 11);

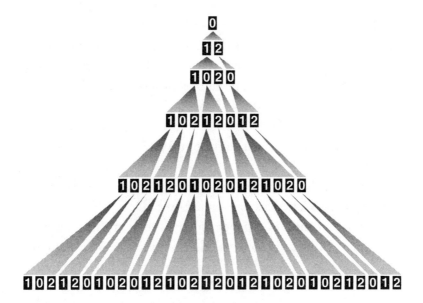

A ternary sequence devised by Axel Thue can be extended indefinitely without ever generating two adjacent identical subsequences of any length. No binary sequence has this property. The sequence is defined by three replacement rules: $0 \rightarrow 12$, $1 \rightarrow 102$, $2 \rightarrow 0$.

then, among sequences three bits long, there are 010 and 101, but none of the other six possibilities is square free. If you now try to create a four-digit square-free binary sequence, you'll find that you're stuck. No such sequences exist.

What about square-free *ternary* sequences? Try to grow one digit by digit, and you're likely to find your path blocked at some point. For example, you might stumble onto the sequence 0102010, which is square free but cannot be extended without creating a square. (Try it!) Many other ternary sequences also lead to such dead ends. Nevertheless, the Norwegian mathematician Axel Thue proved almost a century ago that unbounded square-free ternary sequences exist, and he gave a method for constructing one. The heart of the algorithm is a set of digit replacement rules: $0 \rightarrow 12$, $1 \rightarrow 102$, $2 \rightarrow 0$. At each stage in the construction of the sequence, the appropriate rule is applied to each digit, and the result becomes the starting point for the next stage. The illustration on the previous page shows a few iterations of this process. Thue proved that if you start with a square-free sequence and keep applying the rules, the sequence will grow without bound and will never contain a square.

More recently, attention has turned to the question of how many ternary sequences are square free. Doron Zeilberger of Rutgers University, in a paper co-authored with his computer Shalosh B. Ekhad, established that among the 3^n n-digit ternary sequences, at least $2^{n/17}$ are square free. Uwe Grimm of the Universiteit van Amsterdam has tightened this lower bound somewhat; he has also found an upper bound and has counted all the n-digit sequences up to $n = 110$. It turns out there are 50,499,301,907,904 ways of arranging 110 ternary digits that avoid all repeated patterns. I'll have to choose one of them when I set up my square-free file drawer.

AFTERTHOUGHTS

Not long after this essay appeared, I learned of a remarkable ternary calculating machine built by Thomas Fowler in 1840. Fowler lived in the town of Great Torrington, in Devon, in the southwest of England, far from centers of learning and industry. He had little schooling, and in his youth he was apprenticed to a fellmonger—the tradesman who receives animal hides from the slaughterhouse and prepares them for the tannery. Fowler had other aims; he went on to become a printer and a banker, and also an inventor. In 1835 he was appointed treasurer of the Great Torrington Poor Law Union, a public agency that administered what we would now call a workers' compensation and disability fund. Calculating assessments and payments involved a lot of arithmetic, a subject on which Fowler had innovative ideas. In 1838 he published a book titled *Tables for Facilitating Arithmetical Calculations*, which included methods of computing with both binary and signed ternary numbers. Two years later he built his calculating machine based on the signed ternary system.

Nothing survives of Fowler's machine, not even drawings. But he demonstrated its operation to a delegation from the Royal Society, including Charles Babbage (whose own calculating engine used decimal arithmetic) and Augustus De Morgan, a prominent mathematician and logician. De Morgan subsequently wrote a description of the device for the Royal Society, and it was also mentioned by George Airy, the Astronomer Royal. After these notices, however, it seems that Fowler and his work were forgotten for 150 years.

Fowler's calculator was brought to light again through the efforts of David M. Hogan of North Devon College; Pamela Vass, a Devon writer and historian; and Mark Glusker, a mechanical engineer, model maker, and calculator enthusiast in California. Working from De Morgan's notes and other written sources, Glusker built a working ternary calculator, making every effort to reproduce Fowler's original design. The calculator, made entirely of wood, performs multiplication and division. The digits of numbers are loaded into the mechanism by setting sliding rods, each of which has three positions, corresponding to the digit values 1, 0, and −1. Teeth on the rods engage rungs on a pivoting frame to implement the simple multiplication algorithm for signed ternary numbers. The result is read off from another set of sliding rods.

Fowler's 1840 machine reportedly had a capacity of 55 ternary digits, which would give it an enormous range, equivalent to 26 decimal digits. (On the width × radix scale of economy, Fowler's machine scores 165; a decimal device with the same capacity would have a *wr* product of about 262.) Glusker, Hogan, and Vass cite the following passage from one of Fowler's letters to Airy: "I often reflect that had the Ternary instead of the denary Notation been adopted in the Infancy of Society, Machines something like the present would long ere this have been common, as the transition from mental to mechanical calculation would have been so very obvious and simple."

The question of how best to compare the efficiency of ternary, binary, and other bases has provoked some discussion. Ian East of Oxford Brookes University points out that Norbert Wiener addressed the issue in his famous 1948 book, *Cybernetics*. Wiener bases his analysis on a device not too different from the odometer mentioned above, and yet he reaches a different conclusion. He describes a series of dials or scales, each divided into a fixed number of equal-size numbered segments and equipped with a pointer that can indicate one of the segments. The question is whether it's better to have a few dials, each with many numbered segments, or many dials, each with only a few segments. Wiener's answer depends on the observation that when you are trying to determine which of r segments the pointer lies within, you need to make only $r-1$ decisions. The reason is simply that if you have excluded $r-1$ of the possibilities, then the pointer must lie within the only remaining segment. Because of this "none of the above" option, a mathematical analysis that might otherwise have favored ternary notation instead anoints base 2 as the most efficient scheme.

I have struggled long and hard over this passage in *Cybernetics*. I'm still not sure I follow Wiener's argument, but I caught a glimmer of understanding when I read a 1969 article by

Leonard Ornstein of the Mount Sinai School of Medicine. (Ornstein's work was never published, but apparently the article has been widely circulated; it is available on the Internet at citeseer.ist.psu.edu/ornstein69hierarchic.html.) Ornstein takes up the question of the optimal radix in a quite different context: the construction of hierarchical classifications, such as taxonomies of living organisms. Such a classification is a tree-like structure, where successive branchings from trunk to limbs to twigs to leaves represent ever-finer distinctions. A traditional taxonomy (no longer accepted by biologists but still useful as an illustrative example) divides all living things into plant and animal kingdoms; then the animals are split into phyla such as mollusks, arthropods, and vertebrates; the vertebrates in turn are classified as fishes, reptiles, birds, mammals, and so on; the divisions continue on to individual species at the leaf nodes of the tree. Given a specimen to be classified, the taxonomist traverses the tree from the trunk to the appropriate leaf, perhaps deciding first that the organism is an animal rather than a plant, then recognizing it as a vertebrate, and eventually identifying the species.

What does all this have to do with the economy of numbering systems? Ornstein asks what shape the taxonomic tree should have to make the classification process most efficient, in the sense of minimizing the number of comparisons or decisions. At one extreme, a binary tree has only two branches at each level, but as a result there must be many levels. At the other pole, a flattened, horizontal tree puts all the species on one level, with no hierarchy of kingdoms, phyla, and so forth; there are many options to choose among at that one level. Ornstein, adopting the same line of reasoning used to evaluate numbering systems, shows that a tree with e branches per level is optimal; if the number of branches per level must be an integer, then 3 is the most efficient choice.

This analysis seems to support the ternary hypothesis, but Ornstein introduces a further refinement. In performing a classification, the taxonomist compares the specimen with each available category at each level of the tree; at the top level, for example, this process comes down to asking the two questions "Is it a plant?" and "Is it an animal?" However, if there are no other possibilities, then one of these questions is redundant. If it isn't a plant, it *must* be an animal. In general, if there are r categories and every specimen belongs to exactly one of them, then classification requires only $r-1$ comparisons. Taking advantage of this "default" option reduces the cost of a classification by the ratio $(r-1)/r$. For a ternary tree, with $r = 3$, the search effort is reduced to $2/3$ of the original cost; for a binary tree, with $r = 2$, the savings are $1/2$. This difference is enough to overcome the intrinsic advantage of a ternary structure. Here is the basis of the "none of the above" argument that I failed to understand in Wiener's discussion of computer memory.

Does it follow from this analysis that binary numbering systems really are more efficient than ternary? I think the answer depends on details of implementation. In some circumstances, it is surely possible to save a comparison or a decision, and this would give the edge to binary. For example, in reading the state of a device or a signal that has r possible states, it's enough to check $r-1$ of the alternatives. (If a binary signal is not a 0, it must be a 1; if a ternary signal is not a 0 or a 1, it must be a 2.) But I don't see how this trick can be used to advantage in the hardware for storing or representing numbers and other data in a computer memory. Going back to the example of the odometer dial: each wheel must have r marked segments; there's no way to make do with $r-1$. Likewise, in a memory constructed from transistors or other electronic components, binary devices need two distinguishable states and ternary memory cells must have three states.

One last point of clarification on this theme: much of the discussion above gives the impression that base 2 and base 3 are the only integer candidates for the most efficient numerical notation. That's mostly true. If you want to represent the N numbers from 0 through $N-1$, then 3 is the most economical choice for infinitely many values of N. As I noted in the text, there are 8,487 exceptions to this rule where base 2 is superior to base 3. I failed to mention that there is an 8,488th exception. At this single value of N, base 5 offers the most economical representation—better than either base 2 or base 3. I'll leave it as an exercise to find that unique value of N.

In the matter of the Thue sequence—a square-free sequence of ternary digits—Max Hailperin of Gustavus Adolphus College pointed out that the sequence I illustrated is not in fact the one studied by Axel Thue. I gave the correct digit replacement rules, $0 \rightarrow 12$, $1 \rightarrow 102$, $2 \rightarrow 0$, but I started with an initial digit

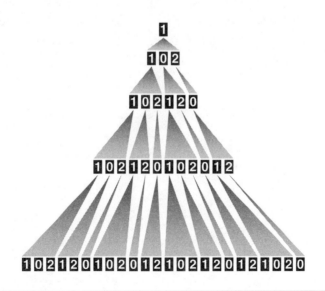

The correct version of the square-free Thue sequence begins with the digit 1 and grows from left to right.

of 0, whereas Thue began with a 1. Both sequences have the square-free property: no pattern of any length appears in two adjacent copies. But Thue's version has the additional interesting property that each stage in generating the sequence is a prefix of the following stage, so that the sequence grows from left to right. The correct Thue sequence is shown in the illustration on the previous page.

Finally, Israel A. Wagner of the IBM Haifa Research Laboratory revealed that the ternary Martha Stewart filing system is even better than I realized. I had remarked that it's possible to insert folders anywhere without ever having two third-cut tabs bump against each other, but removing or deleting a folder spoils this arrangement. Wagner observed that it's always possible to repair the defect by changing the tab of a single folder. For example, suppose the initial state of the file is 0121202 (where the digits 0, 1, and 2 represent the left, middle, and right positions of the tabs). Removing the middle 1 from this sequence brings two 2s together: 012202. But the remedy is simple: either of the 2s can be changed in a way that yields a conflict-free sequence: 010202 or 012102.

CHAPTER 11

Identity Crisis

Suppose I had once borrowed your boat and, secretly, replaced each board with a similar but different one. Then, later, when I brought it back, did I return your boat to you? What kind of question is that? It's really not about boats at all, but about what people mean by "same."

—Marvin Minsky, *The Society of Mind*

I have heard of a touchy owner of a yacht to whom a guest, on first seeing it, remarked, "I thought your yacht was larger than it is"; and the owner replied, "No, my yacht is not larger than it is."

—Bertrand Russell, "On Denoting"

The equal sign seems to be a perfectly innocent bit of mathematical notation. We all know exactly what it means. The symbol = is the fulcrum of a balance. It declares that the things on either side of it, whatever they may be, are equivalent, identical, alike, indistinguishable, the same. What could be clearer? Although an equation may be full of mystery—I can't explain what $e^{i\pi} = -1$ really means—the enigma lies in the two objects being weighed in the balance; the equal sign between them appears to be totally straightforward.

But equality isn't as easy as it looks. Sometimes it's not at all obvious whether two things are equal—or even whether they are two things. In everyday life the subtle ambiguities of identity and equality are seldom noticed because we make unconscious allowances and adjustments for them. In mathematics they cause a little more trouble, but the place where equality gets really queer is in the discrete, deterministic, and literal-minded little world of the digital computer. There, the simple act of saying that two things are "the same" can lead into surprisingly treacherous territory.

What follows is a miscellaneous collection of problems and observations connected in one way or another with the concepts of equality and identity. Some of them are mere quibbles over the meaning of words and symbols, but a few reflect deeper questions. The difficulty of defining equality inside the computer may even shed a bit of light on the nature of identity in the physical world we think we live in.

Some Are More Equal Than Others

There's an old story about the mathematician who sets out to learn a computer programming language such as FORTRAN or C. Everything goes swimmingly until she comes to the statement $x = x + 1$, whereupon she concludes that computer programming is mathematical nonsense.

Of course the story is just a programmer's joke at the expense of mathematicians. I would respond in the same spirit by suggesting that a mathematician would be better equipped than anyone else to solve the "equation" $x = x + 1$. Obviously x is equal to $\aleph_0$, the first infinite cardinal number, which has just the property that $\aleph_0 = \aleph_0 + 1$. (The symbol $\aleph$ is aleph, the first letter of the Hebrew alephbet.)

In truth, $x = x + 1$ is not an equation at all in FORTRAN or C, because the symbol = is not an equal sign in those languages. It is not a relational operator, comparing two quantities, but an assignment operator, which *manufactures* equality. When an assignment statement is executed, whatever is on the left of the = sign is altered to make it equal to the value of the expression on the right. The semantics of this operation are altogether different from those of testing two things for equality. As it happens, the semantics of assignment introduce certain troublesome characteristics into computer programs, which I shall have occasion to mention again below.

To avoid confusion between equality and assignment, many programming languages choose different symbols for the two operations. For example, Algol and its descendants write ":=" for assignment. Throughout this essay, the symbol = will mean only equality (whatever that means).

By the way, the = notation was invented by Robert Recorde (1510–58). He chose two parallel lines as a symbol of mathematical equality "because noe 2 thynges can be moare equalle."

Equality of Beans

Just what does it mean for two numbers to be equal? Note that this is a question about *numbers*, not *numerals*, which are representations of numbers. The numeral 5 in base 10 and the numeral 11 in base 4 and the Roman numeral V all represent the same number; they are all equal. Numbers can even be represented by piles of beans, which form a unary, or base-1, notation.

For small enough hills of beans, most people can judge equality at a glance, but computers are no good at glancing. If you want to teach a computer to test bean piles for equality, you'll need an algorithm. There's a very simple one that works on piles of any finite size. The only mathematical skill the computer needs is the ability to count from 0 up to 1: it has to be

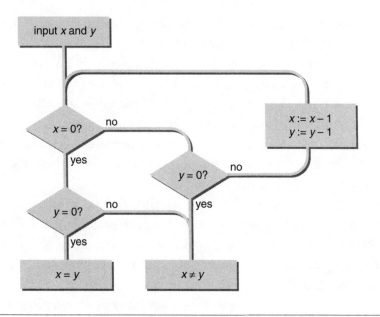

Peano's algorithm establishes the equality of two natural numbers, or nonnegative integers. The key idea is to repeatedly reduce both numbers by 1; if they reach 0 at the same time, they are equal.

able to recognize an empty pile and to choose a single bean from a nonempty pile. Then it can determine the equality of two piles by following these rules: First, check the piles to see if they are empty. If both piles are empty, they are obviously equal. If one pile is empty and the other isn't, the piles are unequal. If neither pile is empty, remove one bean from each pile and repeat the whole procedure. Since the piles are finite, at least one of them must eventually be emptied, and so the algorithm will always terminate.

This method is loosely based on a scheme devised a century ago by the Italian mathematician Giuseppe Peano, who formulated a set of axioms for arithmetic in the natural numbers (also known as the counting numbers, or the nonnegative integers).

Peano's method can be generalized, though awkwardly, to the set of *all* integers, including negative whole numbers. It can even be extended to the rational numbers, which are ratios of whole numbers, as in $1/2$ or $22/7$. Since every rational number can be represented as a fraction reduced to lowest terms, two rationals are equal if their numerators are equal and their denominators are equal. Thus we merely have to apply Peano's algorithm twice to test rationals for equality.

Where numerical equality tests get sticky is with the *irrationals*—numbers that cannot be represented as ratios of integers. A few of these numbers, such as π and e and the square root of 2, have attained celebrity status. They are usually written as decimal approximations, as with the familiar 3.14159 for the first few digits of π. Peano's algorithm won't work with such numbers, because there is no fixed unit you can subtract repeatedly in order to reduce the number to zero. You might think you could compare two irrationals by matching their decimal digits one by one, from left to right. And indeed this scheme will work to prove *in*equality: if two numbers differ, then sooner or later you'll find two digits that fail to match. In the case where the numbers *are* equal, however, the test never ends, because there are always more digits to check.

Comparing numbers written in decimal form has another pitfall as well, which can trip you up even in the case of rational numbers. The problem is that a single number can have multiple decimal expansions, which are mathematically equivalent but don't look at all alike. A case in point is the pair of values 0.999... and 1.000... (where the three dots signify an infinitely repeated pattern of digits). If the digits are matched from left to right, these numerals have not a single digit in common, and yet they denote exactly the same value; there is not the least smidgen of difference between them. (If you doubt this, consider that 0.333... + 0.333... + 0.333... = 0.999..., while $\frac{1}{3} + \frac{1}{3} + \frac{1}{3} = 1$.) Thus some numbers look alike but can't be proved equal, whereas others are equal but look very different.

Same Difference

The integers, the rationals, and the irrationals, taken together, make up the continuum of *real* numbers. It's called a continuum because the numbers are packed together along the real number line with no empty spaces between them. No matter how finely you divide the line, every point corresponds to a number. Almost all those numbers, incidentally, are irrational. If you choose a point at random along the number line, the probability of landing on a rational or an integer is zero.

Real numbers are creatures of mathematics, not computer science. Although some programming languages offer a data type named "real," the numbers of this type are quite *un*real. They can't possibly encompass the infinity of numbers on the real line. In most programming languages, reals are approximated by "floating point" numbers, which have a format much like scientific notation. A number such as 6.02×10^{23} is represented internally by storing the two values 6.02 (called the significand) and 23 (the exponent). Only a finite number of bits are reserved for the significand and the exponent, and so the numbers are limited in both precision and range.

One big advantage of floating-point arithmetic is that you never have to wait forever. When floating-point values stand in for reals, questions about equality are always answered promptly. On the other hand, the answers may well be wrong.

In your favorite programming language, calculate the square root of 2 and then square the result. I've just tried this experiment with an old programmable calculator, which reports the answer as 1.999999999. Interpreted as an approximation to the real number 1.999..., this result is not an error. It's just as correct as 2.000000000, which is also an approximation. The problem is that the machine itself generally cannot recognize the equivalence of the two alternative answers. Suppose a program includes the conditional statement:

if $(\sqrt{2}\,)^2 = 2$
   then let there be light
   else annihilate the universe

If this program happens to be running on my old HP-41C, we're all in trouble.

There are alternatives to floating-point arithmetic that avoid these hazards. Symbolic-mathematics systems such as Maple and Mathematica get the right answer by eschewing numerical approximations; in effect, they define the square root of 2 as "the quantity that when squared is equal to 2." A few programming languages provide exact rational arithmetic. And there have been various schemes for calculating with approximations to real numbers that grow more digits on demand, potentially without limit. Nevertheless, most numerical computation is done with conventional floating-point numbers, and a whole subdiscipline of numerical analysis has grown up to teach people how to cope with the errors.

Programmers are sometimes advised not to compare floating-point values for exact equality, but rather to introduce a small quantity of fudge. Instead of computing the relation $x = y$, they are told to base a decision on the expression $|x - y| < \varepsilon$, where the straight brackets denote the absolute-value operation, and ε is a small number that will cover up the imprecision of floating-point arithmetic. This notion of approximate equality is good enough for many purposes, but it has a cost. Equality loses one of its most fundamental properties: it is no longer a transitive relation. In the presence of fudge, you can't count on the basic principle that if $x = y$ and $y = z$, then $x = z$.

Some Are Less Equal Than Others

Numbers aren't the only things that can be equal or unequal. Most programming languages also have equality operators for other simple data objects, such as alphabetic characters; thus $a = a$ but $a \neq b$. (Whether $a = A$ is a matter up for debate.)

Sequences of characters (usually called strings) are also easy to compare. Two strings are equal if they consist of the same characters in the same sequence, which implies the strings also have the same length. Hence an equality operator for strings simply marches through the two strings in parallel, matching up the characters one by one. Certain other data structures, such as arrays, are handled in much the same way.

But one important kind of data structure can be problematic. The most flexible way of organizing data elements is with links, or pointers, from one item to another. For example, the symbols a, b, and c might be linked into the list $a \rightarrow b \rightarrow c \rightarrow nil$, where *nil* is a special value that marks the end of a chain of pointers. Comparing two such structures for equality is straightforward: just trace the two chains of pointers, and if both reach *nil* at the same time without having encountered any discrepancies along the way, they are identical.

The pointer-following algorithm works well enough in most cases, but consider this structure:

$$a \longrightarrow b \rightarrow c$$

An algorithm that attempts to trace the chain of pointers until reaching *nil* will never terminate, and so equality will never be decided. This problem can be solved—the work-around is to lay down a trail of bread crumbs as you go, and stop following the pointers as soon as you recognize a site you've already visited—but the technique is messy.

There's something else inside the computer that's remarkably hard to test for equality: programs. Even in the simplest cases, where the program is the computational equivalent of a mathematical function, proving equality is a challenge. A function is a program that accepts inputs (called the arguments of the function) and computes a value, but does nothing else to alter the state of the computer. The value returned by the function depends only on the arguments, so that if you apply the function to the same arguments repeatedly, it always returns the same value. For example, $f(x) = x^2$ denotes a function of a single argument x; the function's value is the square of x.

A given function could be written as a computer program in many different ways. At the most trivial level, $f(x) = x^2$ might be replaced by $f(y) = y^2$, where the only change is to the name of the variable. Another alternative might be $f(x) = x \times x$, or perhaps $f(x) = \exp(2 \log(x))$, both of which should produce the same result, at least in the ideal world of mathematics. It seems reasonable to say that two such functions are identical if they return the same value when applied to the same argument. But if that criterion were to serve as a test of function equality, you would have to try all possible arguments within the domain of the function. Even when the domain is not infinite, it is often inconveniently large. The alternative to such an "extensional" test of equality is an "intensional" test, which tries to prove that the texts of the two programs have the same meaning. Fabricating such proofs is not impossible—optimizing compilers do it all the time when they substitute a faster sequence of machine instructions for a slower one—but it is hardly a straightforward task.

When you go beyond programs that model mathematical functions to those that can modify the state of the machine, proving the equality of programs is not just hard but undecidable. That is, there is no algorithm that will always yield the right answer when asked whether two arbitrary programs are equivalent. (For a thorough discussion of program equivalence, see Richard Bird's book *Programs and Machines*.)

One and the Same

We seldom notice it, but words such as "equal," "identical," and "the same" conceal a deep ambiguity. Consider these sentences:

On Friday, Alex and Baxter wore the same necktie.
On Friday, Alex and Baxter had the same teacher.

These two instances of "the same" are not at all the same. Unless Alex and Baxter were yoked together at the neck on Friday, they wore two ties; but they had only one teacher. In the first case two things are alike enough to be considered indistinguishable, and in the second case there is just one thing, which is necessarily the same as itself. The two concepts are so thoroughly entangled that it's hard to find words to speak about them. Where confusion is likely, I shall emphasize the distinction with the terms "separate but equal" and "selfsame."

When there's some uncertainty about whether two things are alike or are really just one thing, the usual strategy is to examine them (it?) more closely. If you study the two neckties long enough, you're sure to find some difference between them. Even identical twins are never truly identical; when you get to know them better, you learn that one has a tattoo and the other can't swim. (If all else fails, you can ask them; they know who they are.)

The strategy of looking harder until you spot a difference doesn't work as well inside the computer, where everything is a pattern of bits and separate patterns really can be equal. Bits have no blemishes or dents or distinguishing features.

Another way to decide between the two kinds of sameness is through the rule of physics (or metaphysics) that a single object cannot be in two places at the same time, and two objects cannot occupy the same space at the same time. Thus all you have to do is bring Alex and Baxter together in the same room and check their neckwear.

The computational equivalent of this idea states that two objects are in fact the selfsame object only if they have the same address in memory. Thus two copies of an object can be distinguished, even though they correspond bit for bit, because they have different addresses. This is a practical method in widespread use, and yet it has certain unsatisfactory aspects. In the first place, it assumes that the computer's memory is organized into an array of unique addresses, which is certainly the usual practice but is not the only possibility. Second, letting identity hinge on location means that an object cannot move without changing into something else. This idea that where you live is who you are contradicts everyday experience. It is also a fiction in modern computer systems, where data are constantly shuffled about by mechanisms such as virtual memory, caches, and the storage-management technique called garbage collection; to maintain the continuity of identity, all of these schemes have to fool programs into thinking that objects don't move—a source of subtle bugs.

More of the Same

There is a third way of exploring issues of identity, but it lies outside the realm of nondestructive testing. If Alex spills ketchup on his tie, does Baxter's tie also acquire a stain? If Baxter steps on his teacher's toe, does Alex's teacher have a sore foot? The principle being suggested here is that two things are the selfsame thing if changing one of them changes the other in the same way.

In computing, this process is encountered most often in the unexpected and unpleasant discovery that two variables are "aliases" referring to the same value or object. For example, if the variable designating Alex's grade point average and the variable for Baxter's average both refer to the same location in memory, then any change in one value will also be reflected in the other. This is probably not the desired behavior of the school's grading system.

In principle, deliberate alteration of memory contents could serve as a test of identity: just twiddle the bits of an object and see if the corresponding bits of another object flip. But the test is not foolproof, particularly in a computer with multiple threads of execution. There's always the chance that the same change might be made coincidentally in two different places; two independent ketchup stains are not an impossibility. If coincidence seems too unlikely, consider that there might be a process running whose purpose is to synchronize two variables, checking one of them at frequent intervals and changing the other one to match. Or, conversely, a background process might undo any change to a variable, restoring the original value whenever it is modified. Under these conditions, a bit-flipping identity test might find that an object is not even equal to itself.

The distinction between separate-but-equal objects and a selfsame object can be crucially important. When I make a bank deposit, I'd strongly prefer that the amount be credited to my own selfsame account rather than to the account of someone who is separate-but-equal to me—perhaps someone with the same name and date of birth. The standard practice for maintaining identity in these circumstances is to issue a unique identifying number. These are the numbers that rule so much of modern life, the ones you find on your bank statement, your driver's license, your credit cards. The same technique can be used internally by a computer program to keep track of data objects. For example, the programming language Smalltalk tags all objects with individual serial numbers. (Smalltalk is also notable for having two equality operators: = for separate-but-equal objects and == for selfsame objects.)

Always the Same

A big advantage of the serial-number approach to identity is that things stay the same even as they change. Identity doesn't depend on location or on any combination of attributes. Two bank accounts might have exactly the same balance, but they are different accounts because they have different account numbers. Within a single account, the balance is likely to vary from day to day, but it remains the selfsame account.

This interplay of constancy and change is certainly a familiar feature of human life. The late Dennis Flanagan wrote that the molecules in most of the tissues of the human body have a residence half-life of less than two weeks. Clearly, then, I'm not the man I used to be—and yet I am. Indeed, it is only when this process of continual molecular replacement ceases that "I" will vanish.

In the semantics of programs, the unique identity of objects matters only when things can change. In a programming system without assignment operators or other ways of modifying existing values, the distinction between separate-but-equal things and the selfsame thing is of no consequence. If an object can never change after it is created, then the outcome of a com-

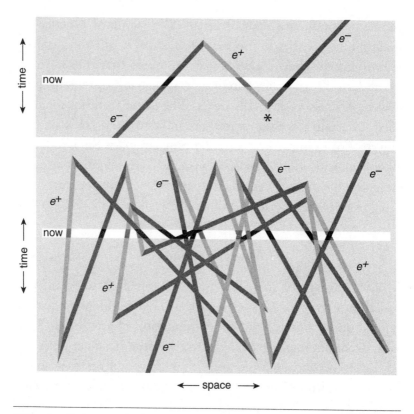

*Why do all electrons look alike? Perhaps they are all the same electron!
The event in the upper diagram is usually interpreted as an interaction
of three particles: an electron (e^-) and a positron (e^+) appear
spontaneously at the point marked with an asterisk, then the positron
collides with a second electron and both particles are annihilated, leaving
only one electron surviving into the future. But the event could also be
the trace of a single particle that moves forward in time as an electron
and backward as a positron. In the lower diagram a single particle with
a tangled world line appears as multiple electrons and positrons.*

putation will never depend on whether the program uses the
original object or an exact copy.

For certain kinds of objects, the whole concept of individual
identity seems beside the point. In the equation $2x - 2 = x + 2$,

should we think of the three 2s as being three separate-but-equal entities, or are they three expressions of a single archetype of 2-ness? It doesn't seem to matter. There is no way of telling one 2 from another. The same can be said of other abstractions, such as alphabetic characters or geometric points.

Even some elements of the physical world share this indifference to individuality. Electrons and other elementary particles seem to be utterly featureless; unlike snowflakes, no two are different. All electrons have exactly the same mass and electric charge, and they carry no serial numbers. They are a faceless multitude. No matter how long and hard we stare, there is no way to tell them apart. They are all separate but equal.

Or else maybe they are all the selfsame electron. In 1948 John Archibald Wheeler, in a telephone conversation with his student Richard Feynman, proposed the delightful hypothesis that there is just one electron in the universe. The single particle shuttles forward and backward in time, weaving a fabulously tangled "world line." At each point where the particle's world line crosses the space-time plane that we perceive as *now*, the particle appears to us as an electron if it is moving forward in time and as a positron (an antielectron) if it is going backward. The sum of all these appearances constructs the material universe. And that's why all electrons have the same mass and charge: because they are all one electron, always equal to itself.

AFTERTHOUGHTS

A classic mathematical conundrum shows how hard it can be to determine equality of real numbers. At issue is the expression $e^{\pi\sqrt{163}}$, and the unlikely-seeming speculation that it might be equal to an integer: 262,537,412,640,768,744. The conjec-ture seems unlikely because three irrational numbers would have to conspire in some extraordinary way to make their com-bination "come out even." On the other hand, there is the fa-mous equation $e^{i\pi} = -1$, already mentioned above, where two irrational numbers (e and π) and an imaginary one (i, or $\sqrt{-1}$) perform an equally remarkable trick. So maybe it's not impos-sible after all?

If you try to confirm or refute this conjecture with an ordi-nary calculator, you won't make much progress; the numbers are too large to be represented exactly in the calculator's mem-ory or display. You'll need a computing system capable of higher-precision calculations. In an amiable 1981 article Philip J. Davis of Brown University makes a series of efforts to settle the question by brute-force computation. Grinding out twenty decimal digits yields the result 262,537,412,640,768,743.99, which is intriguing but inconclusive. Extending the calculation to twenty-five digits gives 262,537,412,640,768,743.9999999, within a ten-millionth of an integer value. But running the pro-gram long enough to generate several more digits finally breaks the pattern: 262,537,412,640,768,743.999999999999250.

If working with real numbers is difficult, floating-point ap-proximations to them are even more treacherous. Neville Holmes of the University of Tasmania pointed out that my strategy of introducing a fudge factor ε in comparisons of floating-point values is not enough to solve the problem. Be-cause the precision of floating-point numbers varies with their size, ε needs to be made a function of the magnitude of the numbers being compared. David W. Cantrell mentioned an-other problem: even the value of 0 is ambiguous in floating-point notation. The prevailing standard for floating-point arithmetic specifies two values of 0, +0.0 and −0.0. According to the standard, the two 0s are equal but nonetheless distinct.

Their reciprocals are as unequal as numbers can be: $+\infty$ and $-\infty$. (Of course one might object that the moment you start taking the reciprocal of 0, you're already in trouble.)

On a lighter note, Les Baxter of Bell Laboratories offered a definitive explanation of the ketchup spill on Alex's tie that caused a stain on Baxter's tie. Les's nephew is Alex Baxter.

CHAPTER 12

Group Theory in the Bedroom

Having run out of sheep the other night, I found myself counting the ways to flip a mattress. Earlier that day I had flipped the very mattress on which I was not sleeping, and the chore had left a residue of puzzled discontent. If you're going to bother at all with such a fussbudget bit of housekeeping, it seems like you ought to do it right, rotating the mattress to a different position each time, so as to pound down the lumps and fill in the sags on all the various surfaces. The trouble is, in the long interval between flips I always forget which way I flipped it last time. Lying awake that night, I was turning the problem over in my head, searching for a golden rule of mattress flipping.

The essential characteristic of a golden rule is universality: one rule works all the time, for everyone. The famous archetype of such rules—the one about doing unto others—certainly has this property. So does the rule of the road: "Drive to the right." ("Drive to the left" works just as well; what matters is not which side you choose but that everyone makes the same choice.) Not all rules generalize so smoothly. A sign saying "Please Use Other Door" is not helpful when it's posted on *every* door. A golden rule of mattress flipping would be some set of geometric maneuvers that you could perform in the same way every time in order to cycle through all the configurations of the mattress. Following this algorithm might entail extra physical labor on each occasion—perhaps making multiple flips or turns—but at least it would eliminate the mental effort of remembering.

If you, too, spend sleepless nights fretting over this problem, I'm afraid I have some disappointing news. You will not find the golden rule of mattress flipping in this essay. As a matter of fact, no such rule exists—at least not in the form I originally imagined. But please read on anyway: the search for a mattress-flipping algorithm leads to some diverting mathematics, not just in the bedroom but also in the garage and at the breakfast table. Furthermore, although I can offer no golden rule for mattress flipping, I do have some practical advice.

Sleepers, Awake

In the morning, when I Googled "mattress flipping," I learned that I'm not the only one who's been obsessing about this silly business. Linda Cobb, The Queen of Clean®, recommends flipping on a seasonal schedule—side to side in spring and fall, and end over end in summer and winter. Or maybe it's the other way around; I forget. A Web site called eHow, which promises "Clear Instructions on How To Do (just about) Everything," offers the following counsel: "Rotate your mattress twice a year, or more often if instructed by the manufacturer. Flip it over completely after the first six months. Then, after another six months, flip it over and turn it so that the head is at the foot of the bed." Is that clear? What would it mean to flip it over *in*completely? And what's the difference, exactly, between rotating, flipping, and turning? Does the final instruction to flip and turn do anything that couldn't be achieved with a single motion?

Another Web page, Phyl's Furniture Facts, takes on the task of defining some of this terminology: "Flipping means to turn it over while rotating means to make a ¼ turn of the mattress while it lies flat on the bed." (I tried the quarter turn, but it didn't look very comfortable.)

Versions of the illustration reproduced below appear on dozens of Web sites. When I first saw this diagram, I thought for a fleeting moment that I had found my golden rule. A quarter turn, a flip, another quarter turn—"AND THERE YOU

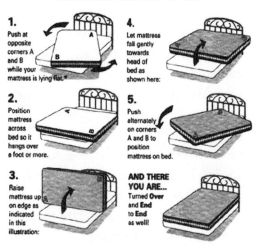

It's easy to turn your mattress properly!
Turn it over and end -to- end.

1. Push at opposite corners A and B while your mattress is lying flat.

2. Position mattress across bed so it hangs over a foot or more.

3. Raise mattress up on edge as indicated in this illustration:

4. Let mattress fall gently towards head of bed as shown here:

5. Push alternately on corners A and B to position mattress on bed.

AND THERE YOU ARE... Turned Over and End to End as well!

TURNING A MATTRESS IS A JOB FOR TWO PEOPLE
Don't risk damage to the mattress or personal injury by doing it yourself.

The mattress-flipping method recommended by several manufacturers and retailers advises you to "turn it over and end-to-end," suggesting that the maneuver rotates the mattress around two axes. In fact the algorithm has the same effect as a single end-over-end flip. (Reprinted from the Natural Bedroom pamphlet by permission of Vivètique.)

ARE . . . Turned Over and End to End as well!" Maybe this was the magic formula. But no: a quick experiment with a small model of a mattress—I used a paperback book—showed that the elaborate sequence of operations in the diagram has exactly the same effect as a half turn end over end. (But the more complicated procedure may be worth following anyway, especially in a room with a low ceiling.)

A Flying Mattress Ride

To make sense of all this turning and flipping, the first thing we need is some clear notation. A mattress can be rotated around any of three orthogonal axes. I could label the axes x, y, and z, but I'd just forget which is which, so it seems better to adopt the terminology of aviation. If you think of a mattress as an airplane flying toward the headboard of the bed, then the three axes are designated roll, pitch, and yaw. The roll axis is parallel to the longest dimension of the mattress (from head to foot), the pitch axis runs along the next-longest dimension (from side to side), and the yaw axis passes through the shortest dimension (top to bottom).

Turning the mattress by 180 degrees around any one of these three axes is a symmetry operation: if you start with the mattress properly installed on the bed and then apply one of these actions, you return to another state where the mattress fits the bed frame correctly. Assuming that the various surfaces of the mattress are not labeled in any way, the states before and after the symmetry operation are indistinguishable. Note that no rotation through an angle smaller than a half turn has this property; despite the advice of Phyl's Furniture Facts, a quarter turn around any axis leaves the mattress in a decidedly awkward position. And for a mattress that has the usual rectangular shape (technically, it's called an *orthotope*), there are no other symmetry axes. If you were to try making a half turn around one of the diagonals, you'd be left with a very cattywampus bed.

A mattress has two sides suitable for sleeping on, and each of those sides has two possible orientations—with one end or the other toward the headboard. Thus there are four configurations overall. A golden rule of mattress flipping would be an operation that, when applied repeatedly, would cycle through all four configurations and then return to the original state. It's easy to see that none of the three basic symmetry operations, taken alone, accomplishes this trick. If you always flip the mattress end over end (that is, around the pitch axis), you alternate between just two of the four states and never reach the other two. Repeated roll turns or yaw turns also visit just two of the states (although not the same pairs of states). Since no single move suffices, any golden rule would have to involve some combination of motions—maybe a roll followed by a pitch followed by a roll the other way and then a yaw, or some such intricate dance move.

Mattress Multiplication

If you are inclined to undertake a search for such a magic combination, by all means flip away; but it's less strenuous to bring some mathematics to bear on the problem. Particularly helpful is the branch of mathematics known as group theory, which is the traditional tool for studies of symmetry.

In general, a group is a set of objects together with an operation for combining them. For example, one can have a group in which the objects are numbers and the combining operation is addition or multiplication. In the case of mattress flipping, however, the "objects"—the elements of the group—are themselves operations. They are the various ways of turning the

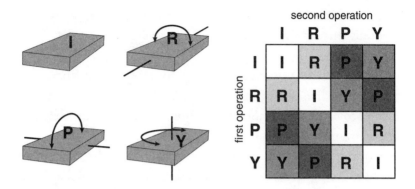

All possible turns and flips of a mattress are cataloged in a mathematical structure called the Klein 4-group. The four allowed motions are the identity (I)—*which is actually a non-motion, leaving the mattress undisturbed—and 180-degree rotations around the roll* (R), *pitch* (P), *and yaw* (Y) *axes. Only these motions restore the mattress to one of the four canonical positions where it is aligned properly with the bed frame. The "multiplication table" for the Klein 4-group, at right, shows the effect of combining any two of the operations. For example, the intersection of the second row and the third column indicates that a roll turn followed by a pitch turn is equivalent to a yaw turn. Each of the operations is its own inverse: doing the same thing twice in a row is the same as doing nothing.*

mattress. The rule for combining the elements is simply to perform one operation after another.

Not just any set of operations can qualify as a group; they have to meet four criteria. First, among the operations there must be an identity element—an operation that leaves the system unchanged. For mattress flipping, the identity operation is obvious: just do nothing.

Second, every element of the group must have an inverse, an "undo" action that returns the system to its former state. Again, this requirement is easily met for mattress flipping: each of the three basic rotations is its own inverse. If you flip the mattress a half turn around the roll axis, and then do exactly the same thing again, you come back to where you started. The same is true of half turns around the pitch and the yaw axes. (Even more obviously, the do-nothing identity element is also its own inverse.)

The third criterion for grouphood is that the operations obey an associative law, so that $(fg)\,h$ means the same as $f\,(g\,h)$, where f, g, and h are any operations in the group. For mattress flipping this is true but not very informative, so I'll say nothing more about it.

The final requirement is closure, which says the set of operations is in some sense complete. More formally, if f and g are any elements of the group, then the combination of f followed by g must also be an operation in the group. The implications of closure are made clear by drawing up a "multiplication table" for the group, as in the illustration on the opposite page. The table gives the result of all possible pairwise combinations of the four operations I, R, P, and Y (which stand for the identity operation and for 180-degree rotations around the roll, pitch, and yaw axes). The crucial point is that every such combination is equivalent to one of the fundamental operations. For example, a roll turn followed by a yaw turn has exactly the same effect as a pitch turn. This fact is what dooms the search for a golden rule. The table shows that any combination of two basic operations can be replaced by a single operation. Since we already know that none of the single operations yields a cycle through all four states of the mattress, it's clear that no pair of operations can be composed to make a golden rule.

What about longer sequences of symmetry moves? They, too, are ruled out. Someone might come forward to claim that a complicated, n-step sequence of roll, pitch, and yaw turns has the desired effect. But by consulting the multiplication table, we can replace the first two of these actions with a single motion, creating an equivalent procedure with $n-1$ steps. Continuing in the same way, we eventually reduce the entire sequence to a single symmetry operation, which we already know cannot be a golden rule.

But wait! Maybe there's some tricky maneuver that involves motions other than the symmetry operations, such as quarter turns or rotations around a diagonal. The trouble is, any move that qualifies as a mattress flip has to begin and end with the mattress in one of the four canonical positions on the bed frame. In between, you are welcome to twirl it over your head on one finger while riding a unicycle, but when you put it down again, only the net effect of your gyrations can be observed. And the multiplication table for the group says that all your manipulations, no matter how acrobatic, can be replaced by a single symmetry operation—either I, R, P, or Y.

There is no golden rule hidden under the mattress.

Group Theory in the Garage

Not all household chores are as mathematically intractable as mattress flipping. In rotating the tires of a car, for example, it's easy to find a golden rule. One simple strategy says: always rotate a quarter turn clockwise. In other words, move the right-front tire to the right rear, move the right-rear tire to the left rear, and so on. The analogous counterclockwise rule works just as well. In either case, when you repeat the process four times, the tires return to their original positions, each one having visited all four corners.

Why is tire rotation so different from mattress flipping? It is governed by a different group. The quarter-turn operations are elements of a group known as the cyclic 4-group, which describes the symmetries of a square that rotates in a plane (but cannot be lifted out of the plane and flipped over). The

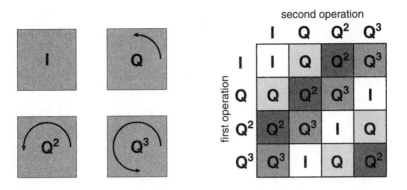

The cyclic 4-group, representing the symmetries of a square confined to a plane, is the only other group of four elements. The elements are the identity (I) and rotations of 90 degrees (Q), 180 degrees (Q^2), and 270 degrees (Q^3); this last motion could also be described as a rotation of 90 degrees in the opposite direction. The notation Q, Q^2, Q^3 is meant to suggest a quarter turn done once, twice, or thrice. The group is called the cyclic 4-group because applying the Q operation four times in a row cycles through all four orientations of the square before returning to the initial state; Q^3 also has this property. In contrast, the Klein 4-group has no single operation that can be performed repeatedly to cycle through all the states.

fundamental symmetry operations are turns of 0 degrees (the identity element), 90 degrees, 180 degrees, and 270 degrees. (Alternatively, the 270-degree turn could be described as a 90-degree rotation in the opposite direction.) The multiplication table for the cyclic 4-group is shown in the illustration above. The quarter-turn and three-quarter-turn operations are golden-rule moves within this group: repeatedly applying either of them cycles through all the orientations of the square.

The other group we have been discussing, the one associated with mattress flipping, is known as the Klein 4-group, after the German mathematician Felix Klein. This group describes the symmetries of a rectangular object rather than a square; more-

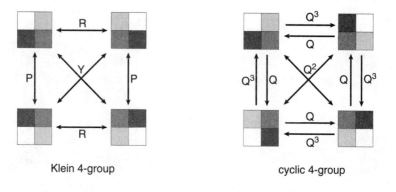

Klein 4-group   cyclic 4-group

Transformations induced by the two four-element groups look very similar on first glance, but subtle differences explain why mattress flipping is harder than certain other chores, such as rotating the tires of a car. Both groups apply to a system with four possible states, represented here by the orientations of a square. But the states in the two diagrams are not the same four states; for example, the configuration at the upper right in the Klein 4-group does not appear anywhere in the cyclic 4-group. The operations Q and Q^3 in the cyclic 4-group are golden rules: to reach every state, you have only to apply a single rule over and over. The Klein 4-group has no such golden-rule operations; reaching all the states requires a combination of moves.

over, the rectangle resides in three-dimensional space, so that it can be flipped over (or, equivalently, reflected in a mirror).

Comparing the multiplication tables for the two groups reveals an important similarity. Both tables are symmetrical with respect to the main diagonal (running from upper left to lower right). In other words, the symbol at row *j* and column *k* is invariably the same as the symbol at row *k* and column *j*. This implies that any two actions can be performed in either sequence with the same effect. For the mattress, a roll followed by a yaw is the same as a yaw followed by a roll. Groups with this property are said to be commutative, or Abelian, after the Norwegian mathematician Niels Henrik Abel.

It turns out that the cyclic 4-group and the Klein 4-group are the only groups with exactly four elements; there is just no other way to combine four operations and satisfy all the requirements of grouphood. But both of the four-element groups are embedded within a larger group, called S_4, which describes all possible permutations of four things. Taking group theory into the kitchen now, S_4 is the group enumerating the ways to arrange a family of four at the breakfast table. The earliest riser can choose any of the four places; then the next person to get out of bed has three chairs available; the third person has two choices; and the last to arrive must take whatever's left. Thus the number of arrangements is 4 factorial (denoted 4!), equal to $4 \times 3 \times 2 \times 1$, or 24. The 24 elements of the group S_4 are all the ways of reshuffling the seat assignments.

Among the twenty-four permutations, nine are derangements—they leave no person in the same chair. Except for the identity, all the elements of both the Klein 4-group and the cyclic 4-group are derangements. There are four more derangements in S_4, not members of either four-element group. These additional derangements provide further useful patterns for rotating tires—as shown in the illustration on the next page—but they still offer no help for the mattress problem.

A Silver Rule

The absence of a golden rule for mattress flipping is a disappointment, but it does not portend the demise of civilization. We can adapt; we can learn to live with it.

Suppose you flip your mattress at regular intervals, but each time you choose an axis of rotation at random. This is clearly a less-than-optimal algorithm, but how large a penalty does it carry? Under the ideal rotation schedule, each orientation of the mattress would get 25 percent of the wear. A quick com-

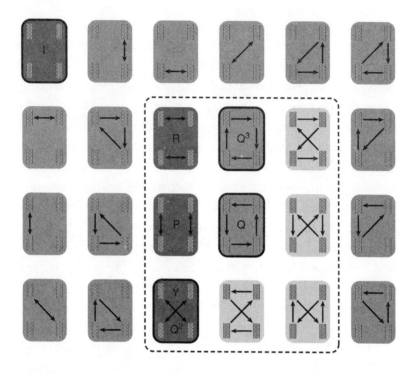

Patterns for rotating the tires of a car are enumerated by the 24-element group called S_4, which lists all possible permutations of 4 objects. Both the Klein 4-group and the cyclic 4-group are subgroups of S_4; here the elements of the Klein group are shown with a dark gray background, and those of the cyclic group have a heavy outline. (Some operations are members of both groups.) The nine operations within the dashed rectangle are derangements, which leave no tire unmoved. The derangements include four more patterns (light gray) that can serve as golden rules for tire rotation, although not for mattress flipping.

puter simulation shows that if you do random flips quarterly over a period of ten years, the most-used orientation will get 31 percent of the wear and the least-used 19 percent. Except for those among us who suffer from a severe princess-and-the-pea complex, ±6 percent is probably good enough.

But we can do even better. We can cheat.

The no-golden-rule theorem for mattress flipping assumes that the surfaces of the mattress are unmarked, so that the four allowed configurations are indistinguishable. Everything changes if you label the surfaces. Specifically, suppose you go through the four possible orientations of the mattress and label each one with a number from the set {0, 1, 2, 3}. You might place the labels so that in each configuration, one of these numbers is facing upward in the corner closest to the right-hand side of the headboard. Given this labeling, the mattress-flipping algorithm calls for nothing more than counting. Each

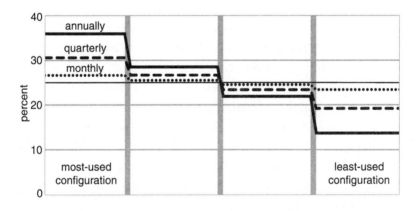

For the laid-back mattress flipper, the simplest algorithm may suffice: don't try to remember which way you did it last time; just pick an axis at random. The graph shows the result of simulating ten years of mattress use under this protocol. Ideally, each of the four configurations would get 25 percent of the wear. If you make a random flip once a year, the mattress winds up in the most-used configuration 36 percent of the time and in the least-used position 14 percent. Doing quarterly flips reduces the imbalance to 31 percent versus 19 percent. A monthly flipping schedule yields an even smaller discrepancy. (But if you are compulsive enough to turn your mattress every month, you probably remember all the details from one flip to the next.)

time you are ready to make a flip, you note the number that appears in the upper-right-hand corner and calculate the successor of that number modulo 4. (In other words, you cycle through the sequence 0, 1, 2, 3 and then return to 0 again.) Turn the mattress in whatever way is necessary to bring the successor number into the upper-right position. The turn needed will not always be around the same axis, but the closure property of the group guarantees that you will always be able to bring the next number into position with a single flip around *some* axis.

The counting algorithm is not a golden rule, but perhaps it deserves to be called a silver one. As a practical matter, this solution is so simple that I would expect mattress makers to adopt it, by embossing numbers on their products. Some of the manufacturers require periodic flipping as a condition of maintaining a warranty, and they give complicated instructions on how to comply. Wouldn't it be easier just to count?

The algorithm can be simplified even further for those who can't count as high as three. If a mattress had lengthwise stripes on one side and crosswise stripes on the other, you could cycle through the four states by always flipping parallel to the stripes. Another possibility is to somehow adapt to our purposes the label that reads, "Do not remove this label." (Now there's a golden rule!)

Pillow Talk

I'm not a mathematician, but I've been hanging around with some of them long enough to know how the game is played. Once you've solved a problem, the next step is to generalize it beyond all recognition. There's a nerdy joke with the punch line "Consider a spherical cow"; in the same spirit I propose, "Consider a cubical mattress."

Consider a cubical mattress: If you cut out the pattern above and fold it to form a cube, the twenty-four possible states (four orientations for each of six faces) are numbered from 0 through 23. Hold the cube so that the face with the smallest numbers is on top and the 0 is in the upper-right corner; then you can proceed through the numbers in sequence by making a series of quarter turns around the three axes.

A cube has a much higher order of symmetry than the generic orthotope of an ordinary mattress. Any of six faces can be turned uppermost for sleeping on, and each face has four orientations, so there are twenty-four states in all. (The group is S_4, the same group we encountered at the breakfast table and in the garage.) Is there a golden rule for flipping an unlabeled cubical mattress—a single maneuver that can be repeated twenty-four times to cycle through all the configurations? The answer is no, but I'll leave the proof of that fact as an exercise.

There *is* a silver rule for cubical mattress flipping: as with an ordinary mattress, if we label the configurations, we can step through them one by one by counting. But there is a subtle difference. On the ordinary mattress, there are six essentially different ways to label the configurations—six ways to arrange the numbers 0 through 3—but it doesn't matter which one you choose. With any arrangement of the numbers, you can always go from one state to the next in a single flip. For the cube, there are many more labelings (the exact number is 23!, which multiplies out to 25,852,016,738,884,976,640,000), and they are not all equal. When you tour all twenty-four states in numerical order, some labelings allow every transition to be accomplished by a simple quarter turn around the roll, the pitch, or the yaw axis. Other labelings require more complicated maneuvers, such as rotations around multiple axes. How many of the labelings fall into each category? Is there any simple rule that distinguishes the two classes?

If you have answered those questions and you still can't get to sleep, you are welcome to go on and consider a hyper-cubical mattress—one that takes the form of a cube in n-dimensional space, for some n greater than three. The four-dimensional mattress would have twenty-four square faces.

As far as I know, neither Sealy nor Simmons nor anyone else is yet selling a four-dimensional mattress. As a matter of fact, the trend seems to be in the opposite direction. Mattress makers now promote the one-sided "no-flip" mattress, whose only symmetry operation is a 180-degree turn around the yaw axis. This innovation finally gives us a golden rule for mattress flipping, but it solves the problem in a disappointing and unsatisfying way.

Still, there remain ways to increase the number of permutations, to provide opportunities for creative problem solving, to make life more interesting. One could get married.

AFTERTHOUGHTS

There's a subject I didn't get a chance to mention when this essay was first published: the mattress business is really weird. In looking for advice on mattress flipping, I found myself reading sales brochures and shopping guides for bedding—a literature I have seldom had occasion to consult. One of the strangest features of mattress marketing is the ambivalent attitude toward firmness. Apparently, the ideal mattress is both rock hard and feather soft. Seth Stevenson, writing in the online magazine *Slate*, reports finding a mattress called the Simmons Beautyrest World Class Granite Plush.

In the course of my research, I discovered that one chain of mattress stores offers an e-mail reminder service for forgetful flippers. When you buy a mattress from the Original Mattress Factory, you are invited to enroll in the Mattress-Flip Reminder System. The truth is, I didn't buy one of their mattresses, but I *did* sign up for the reminders, and I've been receiving quarterly e-mails ever since with the subject line: "It's time to Flip/Rotate your mattress!" What amuses and appalls me about these notices is that they offer no help at all in remembering *which way* to Flip/Rotate.

A few readers reported seeing mattresses that came premarked in one way or another for a silver-rule flipping routine. Many others offered helpful do-it-yourself advice on the same theme, involving Magic Markers, Dymo tape, Post-it notes, and such. There were also mnemonics, such as "Spin in the spring, flip in the fall" and "Rodd and Yeven" (meaning to roll in odd-numbered months and yaw in even ones).

Michael Malcolm of Sunnyvale, California, suggested a refinement to my own idea of numbering the orientations from 0 through 3 and then always performing whatever movement will bring the next number in sequence to the upper-right corner. I had thought that all of the six possible numberings of the faces were equally good for this purpose, but Malcolm pointed out that some of the numberings are more idiotproof than others. In any legal configuration of the mattress, two numbers are visible on the upper face—one at the upper-right corner and the other at the lower-left. What if I should forget which of the corners is the counting corner? If the arrangement of numbers puts either 0 and 1 or 0 and 3 on the same face, then through pathologically bad memory I could wind up repeating a yaw rotation over and over again, and I would never see the opposite face. But the two numberings that bring 0 and 2 together on the same face (putting 1 and 3 on the opposite face) have a marvelous property: it doesn't matter which of the visible numbers you choose as your counting cue! Whether or not you are consistent in this choice, you will always alternate roll and pitch turns (and never perform a yaw turn). In fact, these two numberings are equivalent to marking the two faces with stripes or arrows indicating the roll and pitch axes.

My friend Barry Cipra of Northfield, Minnesota, proposed: "Consider a pentagonal mattress." If you sleep on a regular pentagon, there are ten distinguishable states, but again there is no golden rule for cycling through them. Cipra suggests a silver rule: First, mark one side of the mattress with a clockwise circular arrow, and the other side with a counterclockwise arrow. Now, each time the mattress is due for a turn, flip it around the roll axis (which runs from the midpoint of the headboard to the opposite vertex), and then turn it around the yaw axis one-fifth of a revolution in the direction indicated by the arrow. This scheme will also work for an equilateral triangle or indeed for any regular polygon with an odd number of sides. (For an n-gon, the yaw turn is through $n/360$ degrees.)

Cipra's colleague Paul Zorn of St. Olaf College took this idea to the limit—namely, a circular mattress. Here, at last, Zorn finds a golden rule: First flip the mattress over around any diameter. (This can be seen as either a roll or a pitch maneuver—the distinction has disappeared.) Then rotate around the center point (a yaw turn) by an irrational multiple of 360 degrees. The irrationality of the angle is the key: it guarantees that the mattress will never visit any configuration twice. One might quibble that the original specification for a golden rule called for cycling through all the states and then repeating, whereas Zorn's procedure *never* repeats. Since the circular mattress has infinitely many orientations, however, Zorn's approach seems like a reasonable extrapolation. Another correspondent points out that one special rotation angle produces the maximum separation between successive positions of the mattress. This angle (expressed as a fraction of a full circle) is $1/\varphi$, or about 0.618.

My favorite suggestion for an improved mattress came from Peter M. Lawrence of Melbourne, Australia. It's the Möbius mattress. The topological toy called a Möbius strip is usually made from paper: you put a half twist in a strip and then glue the ends together. Doing this with a mattress would require a good deal more effort, but the result would be worth it. The Möbius strip is a *one-sided* object; it has no front and back or top and bottom; you can reach any point on the surface from any other point without ever crossing an edge. Thus the Möbius mattress never has to be turned over, and there's a simple golden rule for cycling through all the regions of its surface.

Further Reading

Chapter 1: Clock of Ages

Brand, Stewart. 1999. *The Clock of the Long Now: Time and Responsibility.* New York: Basic Books.

Dershowitz, Nachum, and Edward M. Reingold. 1990. Calendrical calculations. *Software: Practice and Experience* 20:899–928.

Hayes, Brian. 1995. Computing science: Waiting for 01-01-00. *American Scientist* 83:12–15.

King, Henry C. 1978. *Geared to the Stars: The Evolution of Planetariums, Orreries, and Astronomical Clocks.* In collaboration with John R. Millburn. Toronto: University of Toronto Press.

Knuth, Donald E. 1962. The calculation of Easter . . . *Communications of the Association for Computing Machinery* 5:209–10.

———. 1973. *The Art of Computer Programming.* Vol. 1, *Fundamental Algorithms.* Reading, Mass.: Addison-Wesley. (Chapter 1.3.2, Exercise 14, pp. 155–56; answers to exercises pp. 511–13.)

Lehni, Roger. 1992. *Strasbourg Cathedral's Astronomical Clock.* Translated by R. Beaumont-Craggs. Paris: Editions La Goélette.

Lemley, Brad. 2005. Time machine. *Discover* 26(11):29–35.

Lloyd, H. Alan. 1958. *Some Outstanding Clocks over Seven Hundred Years, 1250–1950.* London: Leonard Hill Limited.

Maurice, Klaus, and Otto Mayr, eds. 1980. *The Clockwork Universe: German Clocks and Automata, 1550–1650*. New York: Neale Watson Academic Publications.

Mayr, Otto. 1986. *Authority, Liberty, and Automatic Machinery in Early Modern Europe*. Baltimore: Johns Hopkins University Press.

Reese, Ronald Lane, Steven M. Everett, and Edwin D. Craun. 1981. The origin of the Julian period: An application of congruences and the Chinese remainder theorem. *American Journal of Physics* 49:658–61.

Reingold, Edward M., and Nachum Dershowitz. 1993. Calendrical calculations, II: Three historical calendars. *Software: Practice and Experience* 23:383–404.

Stansifer, Ryan. 1992. The calculation of Easter. *ACM Sigplan Notices* 27(12):61–65.

Van den Bossche, Benoît. 1997. *Strasbourg: La cathédrale*. Photographs by Claude Sauvageot. Saint-Léger-Vauban: Zodiaque.

Chapter 2: Random Resources

Aumann, Yonatan, Yan Zong Ding, and Michael O. Rabin. 2002. Everlasting security in the bounded storage model. *IEEE Transactions on Information Theory* 48(6):1668–80.

Bauke, Heiko, and Stephan Mertens. 2004. Pseudo random coins show more heads than tails. *Journal of Statistical Physics* 114:1149–69.

Ding, Yan Zong, and Michael O. Rabin. 2002. Hyper-encryption and everlasting security. In *Proceedings of the Nineteenth Annual Symposium on Theoretical Aspects of Computer Science*, pp. 1–26. Lecture Notes in Computer Science, vol. 2285. London: Springer Verlag.

Fisher, R. A., and F. Yates. 1938. *Statistical Tables for Biological, Agricultural, and Medical Research*. London: Oliver and Boyd.

Ford, Joseph. 1983. How random is a coin toss? *Physics Today* 36(4): 40–47.

Marsaglia, George. 1995. The Marsaglia random number CDROM, including the DIEHARD battery of tests of randomness. Tallahassee: Department of Statistics, Florida State University.

Maurer, Ueli M. 1992. Conditionally-perfect secrecy and a provably-secure randomized cipher. *Journal of Cryptology* 5(1):53–66.

Pearson, Karl. 1900. On the criterion that a given system of deviations from the probable in the case of a correlated system of variables is such that it can be reasonably supposed to have arisen from random sampling. *London, Edinburgh, and Dublin Philosophical Magazine and Journal of Science*, series 5, 50:157–75.

RAND Corporation. 1955. *A Million Random Digits with 100,000 Normal Deviates*. Glencoe, Ill.: Free Press.

Shannon, C. E. 1949. Communication theory of secrecy systems. *Bell System Technical Journal* 28:656–715.

Thomson, William [Lord Kelvin]. 1901. Nineteenth century clouds over the dynamical theory of heat and light. *London, Edinburgh, and Dublin Philosophical Magazine and Journal of Science*, series 6, 2:1–40.

Tippett, L.H.C. 1927. Random sampling numbers. In *Tracts for Computers*, no. 15. London: Cambridge University Press.

Tu, Shu-Ju, and Ephraim Fischbach. 2005. A study on the randomness of the digits of π. *International Journal of Modern Physics C* 16(2):281–94.

Vincent, C. H. 1970. The generation of truly random binary numbers. *Journal of Physics E* 3(8):594–98.

Volchan, Sérgio B. 2002. What is a random sequence? *American Mathematical Monthly* 109:46–63.

von Neumann, John. 1951. Various techniques used in connection with random digits. In *Collected Works*, vol. 5, pp. 768–70. New York: Pergamon Press.

Chapter 3: Follow the Money

Angle, John. 1986. The surplus theory of social stratification and the size distribution of personal wealth. *Social Forces* 65:293–326.

———. 1992. The inequality process and the distribution of income to blacks and whites. *Journal of Mathematical Sociology* 17:77–98.

———. 1993. Deriving the size distribution of personal wealth from "the rich get richer, the poor get poorer." *Journal of Mathematical Sociology* 18:27–46.

———. 1996. How the gamma law of income distribution appears invariant under aggregation. *Journal of Mathematical Sociology* 21:325–58.

Blaug, Mark, ed. 1992. *Vilfredo Pareto (1848–1923)*. London: Edward Elgar.

Bouchaud, Jean-Philippe, and Marc Mézard. 2000. Wealth condensation in a simple model of economy. *Physica A* 282:536–45.

Bourguignon, François, and Christian Morrisson. 2002. Inequality among world citizens: 1820–1990. *American Economic Review* 92(4):727–44.

Chakraborti, Anirban. 2002. Distributions of money in model markets of economy. arxiv.org/abs/cond-mat/0205221.

Chakraborti, Anirban, and Bikas K. Chakrabarti. 2000. Statistical mechanics of money: How saving propensity affects its distribution. *European Physical Journal* B17:167–70.

Champernowne, D. G. 1953. A model of income distribution. *Economic Journal* 63:318–51.

———. 1974. A comparison of measures of inequality of income distribution. *Economic Journal* 84:787–816.

Chen, Shaohua, and Martin Ravallion. 2004. How have the world's poorest fared since the early 1980s? World Bank Policy Research Working Paper 3341.

Drăgulescu, A., and V. M. Yakovenko. 2000. Statistical mechanics of money. *European Physical Journal* B17:723–29.

———. 2001. Evidence for the exponential distribution of income in the USA. *European Physical Journal* B20:585–89.

Ispolatov, S., P. L. Krapivsky, and S. Redner. 1998. Wealth distributions in asset exchange models. *European Physical Journal* B2:267–76.

Kadanoff, Leo P. 1971. An examination of Forrester's *Urban Dynamics*. *Simulation* 16:261–68.

Laguna, M. F., S. Risau Gusman, and J. R. Iglesias. 2005. Economic exchanges in a stratified society: End of the middle class? arxiv.org/abs/physics/0505157.

Lambert, Peter J. 1993. *The Distribution and Redistribution of Income: A Mathematical Analysis.* 2nd ed. Manchester, U.K.: Manchester University Press.

Lux, Thomas. 2005. Emergent statistical wealth distributions in simple monetary exchange models: A critical review. arxiv.org/abs/cs.MA/0506092.

Mandelbrot, Benoit B. 1997. *Fractals and Scaling in Finance: Discontinuity, Concentration, Risk.* New York: Springer Verlag.

Mantegna, Rosario N., and H. Eugene Stanley. 2000. *An Introduction to Econophysics: Correlations and Complexity in Finance.* New York: Cambridge University Press.

Montroll, Elliott W., and Wade W. Badger. 1974. *Introduction to Quantitative Aspects of Social Phenomena.* New York: Gordon and Breach.

Pareto, Vilfredo. 1896, 1897, reprinted 1964. *Cours d'économie politique.* Geneva: Librairie Droz.

Ruskin, John. 1862. *Unto This Last: Four Essays on the First Principles of Political Economy.* Edited with an introduction by Lloyd J. Hubenka. Lincoln: University of Nebraska Press.

Sen, Amartya Kumar. 1973, 1997. *On Economic Inequality.* Oxford: Clarendon Press.

Sinha, Sitabhra. 2005. Evidence for power-law tail of the wealth distribution in India. arxiv.org/abs/cond-mat/0502166.

Souma, Wataru. 2002. Physics of personal income. arxiv.org/abs/cond-mat/0202388.

Stanley, H. E., P. Gopikrishnan, V. Plerou, and L.A.N. Amaral. 2000. Quantifying fluctuations in economic systems by adapting methods of statistical physics. *Physica A* 287:339–61.

Chapter 4: Inventing the Genetic Code

Alff-Steinberger, C. 1969. The genetic code and error transmission. *Proceedings of the National Academy of Sciences of the U.S.A.* 64:584–91.

Antoneli, Fernando, Jr., Michael Forger, and José Eduardo M. Hornos. 2004. The search for symmetries in the genetic code: Finite groups. *Modern Physics Letters B* 18:971–78.

Béland, Pierre, and T.F.H. Allen. 1994. The origin and evolution of the genetic code. *Journal of Theoretical Biology* 170:359–65.

Böck, A., K. Forchhammer, J. Heider, W. Leinfelder, G. Sawers, B. Veprek, and F. Zinoni. 1991. Selenocysteine: The 21st amino acid. *Molecular Microbiology* 5:515–20.

Brenner, S. 1957. On the impossibility of all overlapping triplet codes in information transfer from nucleic acid to proteins. *Proceedings of the National Academy of Sciences of the U.S.A.* 43:687–94.

Crick, Francis H. C. 1966. The genetic code—yesterday, today, and tomorrow. In *The Genetic Code, Proceedings of the XXXI Cold Spring Harbor Symposium on Quantitative Biology,* pp. 3–9. Cold Spring Harbor, N.Y.: Cold Spring Harbor Laboratory of Quantitative Biology.

———. 1988. *What Mad Pursuit: A Personal View of Scientific Discovery.* New York: Basic Books.

Crick, Francis H. C., John S. Griffith, and Leslie E. Orgel. 1957. Codes without commas. *Proceedings of the National Academy of Sciences of the U.S.A.* 43:416–21.

Freeland, Stephen J., and Laurence D. Hurst. 1998. The genetic code is one in a million. *Journal of Molecular Evolution* 47:238–48.

Freeland, Stephen J., Tao Wu, and Nick Keulmann. 2003. The case for an error minimizing standard genetic code. *Origins of Life and Evolution of the Biosphere* 33:457–77.

Gamow, George. 1954a. Possible relation between deoxyribonucleic acid and protein structures. *Nature* 173:318.

———. 1954b. Possible mathematical relation between deoxyribonucleic acid and proteins. *Det Kongelige Danske Videnskabernes Selskab, Biologiske Meddelelser* 22:1–13.

Gamow, George, Alexander Rich, and Martynas Yčas. 1956. The problem of information transfer from nucleic acids to proteins. In *Advances in Biological and Medical Physics*, vol. 4, pp. 23–68. New York: Academic Press.

Golomb, S. W. 1962. Efficient coding for the desoxyribonucleic channel. In *Proceedings of Symposia in Applied Mathematics*, vol. 14, Mathematical Problems in the Biological Sciences, pp. 87–100. Providence: American Mathematical Society.

Golomb, S. W., Basil Gordon, and L. R. Welch. 1958. Comma-free codes. *Canadian Journal of Mathematics* 10:202–9.

Golomb, S. W., L. R. Welch, and M. Delbrück. 1958. Construction and properties of comma-free codes. *Det Kongelige Danske Videnskabernes Selskab, Biologiske Meddelelser* 23(9):1–34.

Haig, David, and Laurence D. Hurst. 1991. A quantitative measure of error minimization in the genetic code. *Journal of Molecular Evolution* 33:412–17.

Hayes, Brian. 2004. Ode to the code. *American Scientist* 92:494–98.

Itzkovitz, Shalev, and Uri Alon. 2007. The genetic code is nearly optimal for allowing arbitrary additional information within protein-coding sequences. *Genome Research* DOI:10.1101/gr.5987307.

Jiménez-Montaño, Miguel A., Carlos R. de la Mora-Basáñez, and Thorsten Pöschel. 1995. On the hypercube structure of the genetic code. In *Proceedings of the Third International Conference on Bioinformatics and Genome Research*, ed. Hwa A. Lim and Charles A. Cantor, pp. 445–55. River Edge, N.J.: World Scientific.

Judson, Horace Freeland. 1996. *The Eighth Day of Creation: Makers of the Revolution in Biology.* Expanded ed. Plainview, N.Y.: Cold Spring Harbor Laboratory Press.

Leder, Philip, and Marshall W. Nirenberg. 1964. RNA codewords and protein synthesis, II. Nucleotide sequence of a valine RNA codeword. *Proceedings of the National Academy of Sciences of the U.S.A.* 52:420–27.

Nirenberg, Marshall W., and J. Heinrich Matthaei. 1961. The dependence of cell-free protein synthesis in *E. coli* upon naturally occurring or synthetic polyribonucleotides. *Proceedings of the National Academy of Sciences of the U.S.A.* 47:1588–1602.

Robin, Stephane, François Rodolphe, and Sophie Schbath. 2005. *DNA: Words and Models.* Cambridge, U.K.: Cambridge University Press.

Sella, Guy, and David H. Ardell. 2002. The impact of message mutation on the fitness of a genetic code. *Journal of Molecular Evolution* 54:638–51.

Sinsheimer, Robert L. 1959. Is the nucleic acid message in a two-symbol code? *Journal of Molecular Biology* 1:218–20.

Woese, Carl R. 1967. *The Genetic Code: The Molecular Basis for Genetic Expression.* New York: Harper and Row.

Chapter 5: Statistics of Deadly Quarrels

Ashford, Oliver M. 1985. *Prophet—or Professor? The Life and Work of Lewis Fry Richardson.* Boston: Adam Hilger.

Brecke, Peter. 1999. Violent conflicts 1400 A.D. to the present in different regions of the world. www.inta.gatech.edu/peter/PSS99_paper.html.

Cioffi-Revilla, Claudio A. 1990. *The Scientific Measurement of International Conflict: Handbook of Datasets on Crises and Wars, 1945–1988.* Boulder, Colo.: Lynne Rienner Publishers.

Doyle, Michael W. 1983. Kant, liberal legacies, and foreign affairs. *Philosophy and Public Affairs* 12(3):205–35.

Geller, Daniel S., and J. David Singer. 1998. *Nations at War: A Scientific Study of International Conflict.* Cambridge, U.K.: Cambridge University Press.

Layne, Christopher. 1994. Kant or cant: The myth of the democratic peace. *International Security* 19(2):5–49.

Maoz, Zeev, and Nasrin Abdolali. 1989. Regime types and international conflict, 1816–1976. *Journal of Conflict Resolution* 33:3–35.

Richardson, Lewis Fry. 1960a. *Arms and Insecurity: A Mathematical Study of the Causes and Origins of War.* Edited by Nicolas Rashevsky and Ernesto Trucco. Pittsburgh: Boxwood Press.

———. 1960b. *Statistics of Deadly Quarrels.* Edited by Quincy Wright and C. C. Lienau. Pittsburgh: Boxwood Press.

———. 1961. The problem of contiguity: An appendix to *Statistics of Deadly Quarrels. Yearbook of the Society for General Systems Research*, vol. 6, pp. 140–87. Ann Arbor, Mich.

———. 1993. *Collected Papers of Lewis Fry Richardson*. Edited by Oliver M. Ashford et al. New York: Cambridge University Press.

Richardson, Stephen A. 1957. Lewis Fry Richardson (1881–1953): A personal biography. *Journal of Conflict Resolution* 1:300–304.

Russett, Bruce M., Christopher Layne, David E. Spiro, and Michael W. Doyle. 1995. Correspondence: The democratic peace. *International Security* 19(4):164–84.

Sarkees, Meredith Reid. 2000. The Correlates of War data on war: an update to 1997. *Conflict Management and Peace Science* 18(1):123–44.

Singer, J. David, and Melvin Small. 1972. *The Wages of War, 1816–1965: A Statistical Handbook*. New York: John Wiley.

Sorokin, Pitirim A. 1937. *Social and Cultural Dynamics*. Vol. 3, *Fluctuation of Social Relationships, War, and Revolution*. New York: American Book Company.

Spiro, David E. 1994. The insignificance of the liberal peace. *International Security* 19(2):50–86.

Wilkinson, David. 1980. *Deadly Quarrels: Lewis F. Richardson and the Statistical Study of War*. Berkeley: University of California Press.

Wright, Quincy. 1965. *A Study of War, with a Commentary on War Since 1942*. 2nd ed. Chicago: University of Chicago Press.

Chapter 6: Dividing the Continent

Band, Lawrence E. 1986. Topographic partition of watersheds with digital elevation models. *Water Resources Research* 22(1):15–24.

Beucher, S., and C. Lanteujoul. 1979. Use of watersheds in contour detection. In *International Workshop on Image Processing: Real-Time Edge and Motion Detection/Estimation*, Rennes, France, Sept. 17–21, 1979. cmm.ensmp.fr/~beucher/publi/watershed.pdf.

Cayley, A. 1859. On contour and slope lines. *London, Edinburgh, and Dublin Philosophical Magazine and Journal of Science* 18:264–68. Reprinted in *The Collected Mathematical Papers of Arthur Cayley*, vol. 4, pp. 108–11. Cambridge, U.K.: Cambridge University Press, 1891.

Maxwell, James Clerk. 1870. On hills and dales. *London, Edinburgh, and Dublin Philosophical Magazine and Journal of Science* 40:421–25. Reprinted in *The Scientific Papers of James Clerk Maxwell*, vol. 2, pp. 233–40. New York: Dover Publications.

O'Callaghan, John F., and David M. Mark. 1984. The extraction of drainage networks from digital elevation data. *Computer Vision, Graphics, and Image Processing* 28:323–44.

Vincent, Luc, and Pierre Soille. 1991. Watersheds in digital spaces: An efficient algorithm based on immersion simulations. *IEEE Transactions on Pattern Analysis and Machine Intelligence* 13:583–98.

Chapter 7: On the Teeth of Wheels

Archibald, R. C. 1951. Review of *The Farey Series of Order 1025*, edited by E. H. Neville. *Mathematical Tables and Other Aids to Computation* 5(35):135–39.

Baillie, G. H. 1947. *Watchmakers and Clockmakers of the World*. 2nd ed. London: N.A.G. Press.

Brocot, Achille. 1861. Calcul des rouages par approximation, nouvelle méthode. *Revue chronométrique. Journal des horlogers, scientifique et pratique* 3:186–94.

Bruckheimer, Maxim, and Abraham Arcavi. 1995. Farey series and Pick's area theorem. *Mathematical Intelligencer* 17(4):64–67.

Camus, Charles-Étienne-Louis. 1842. *A Treatise on the Teeth of Wheels, Demonstrating the Best Forms Which Can Be Given to Them for the Purposes of Machinery; Such as Mill-work and Clock-work, and the Art of Finding Their Numbers.* Translated from the French of M. Camus. A new edition, carefully revised and enlarged. With details of the present practice of mill-wrights, engine makers, and other mechanists, by John Isaac Hawkins, civil engineer. London: M. Taylor.

Caruso, Horacio A., and Sebastián M. Marotta. 2000. Vibonacci series of complex numbers. Unpublished manuscript.

Eisenstein, Gotthold. 1850. Letter to M. A. Stern, Jan. 14, 1850. Reprinted in Gotthold Eisenstein, *Mathematische Werke*, vol. 2, pp. 818–23. New York: Chelsea, 1975.

Farey, J. 1816. On a curious property of vulgar fractions. *Philosophical Magazine and Journal* 47:385–86.

Freeth, T., Y. Bitsakis, X. Moussas, J. H. Seiradakis, A. Tselikas, H. Mangou, M. Zafeiropoulou, R. Hadland, D. Bate, A. Ramsey, M. Allen, A. Crawley, P. Hockley, T. Malzbender, D. Gelb, W. Ambrisco, and M. G.

Edmunds. 2006. Decoding the ancient Greek astronomical calculator known as the Antikythera mechanism. *Nature* 444:587–91.

Graham, Ronald L., Donald E. Knuth, and Oren Patashnik. 1989. *Concrete Mathematics: A Foundation for Computer Science.* Reading, Mass.: Addison-Wesley.

Haros. 1802. Tables pour évaluer une fraction ordinaire avec autant de décimales qu'on voudra; et pour trouver la fraction ordinaire la plus simple, et qui approche sensiblement d'une fraction décimale. *Journal de l'École polytechnique,* cahier 11, tome 4, pp. 364–68.

Lagarias, J. C., and C. P. Tresser. 1995. A walk along the branches of the extended Farey tree. *IBM Journal of Research and Development* 39(3):283–94.

Lehmer, D. H. 1929. On Stern's diatomic series. *American Mathematical Monthly* 36(2):59–67.

Merritt, Henry Edward. 1947. *Gear Trains: Including a Brocot Table of Decimal Equivalents and a Table of Factors of All Useful Numbers up to 200,000.* London: Sir Isaac Pitman and Sons.

Pulzer, Peter. 2000. Emancipation and its discontents: The German-Jewish dilemma. Centre for German-Jewish Studies, University of Sussex. www.sussex.ac.uk/Units/cgjs/pubs/rps/RP1B.htm

Stern, M. A. 1858. Ueber eine zahlentheoretische Funktion. *Journal für die reine und angewandte Mathematik* 55:193–220.

Viswanath, Divakar. 1998. Random Fibonacci sequences and the number 1.13198824. *Mathematics of Computation* 69(231):1131–55.

Zeeman, E. C. 1986. Gears from the Greeks. *Proceedings of the Royal Institution of Great Britain* 58:137–56.

Chapter 8: The Easiest Hard Problem

Borgs, Christian, Jennifer T. Chayes, and Boris Pittel. 2001a. Sharp threshold and scaling window for the integer partitioning problem. In *Proceedings of the 2001 ACM Symposium on the Theory of Computing,* pp. 330–36.

———. 2001b. Phase transition and finite-size scaling for the integer partitioning problem. *Random Structures and Algorithms* 19:247–88.

Cheeseman, Peter, Bob Kanefsky, and William M. Taylor. 1991. Where the really hard problems are. In *Proceedings of the International Joint Conference on Artificial Intelligence,* vol. 1, pp. 331–37.

Dubois, O., R. Monasson, B. Selman, and R. Zecchina, eds. 2001. Special issue on phase transitions in combinatorial problems. *Theoretical Computer Science* 265(1).

Erdős, Paul, and A. Rényi. 1960. On the evolution of random graphs. *Publications of the Mathematical Institute of the Hungarian Academy of Sciences* 5:17–61.

Fu, Yaotian. 1989. The use and abuse of statistical mechanics in computational complexity. In *Lectures in the Sciences of Complexity*, ed. Daniel L. Stein, pp. 815–26. Reading, Mass.: Addison-Wesley.

Gent, Ian P., and Toby Walsh. 1996. Phase transitions and annealed theories: Number partitioning as a case study. In *Proceedings of the 1996 European Conference on Artificial Intelligence*, pp. 170–74.

Graham, Ronald L. 1969. Bounds on multiprocessing timing anomalies. *SIAM Journal on Applied Mathematics* 17:416–29.

Karmarkar, Narendra, and Richard M. Karp. 1982. The differencing method of set partitioning. Technical Report UCB/CSD 82/113, Computer Science Division, University of California, Berkeley.

Karmarkar, Narendra, Richard M. Karp, George S. Lueker, and Andrew M. Odlyzko. 1986. Probabilistic analysis of optimum partitioning. *Journal of Applied Probability* 23:626–45.

Mertens, Stephan. 2000. Random costs in combinatorial optimization. *Physical Review Letters* 84:1347–50.

———. 2001. A physicist's approach to number partitioning. *Theoretical Computer Science* 265:79–108.

Chapter 9: Naming Names

Airline Codes Web Site. www.airlinecodes.co.uk

Alter, Adam L., and Daniel M. Oppenheimer. 2006. Predicting short-term stock fluctuations by using processing fluency. *Proceedings of the National Academy of Sciences of the U.S.A.* 103:9369–72.

Book Industry Study Group. 2004. The evolution in product identification: Sunrise 2005 and the ISBN-13. www.bisg.org/docs/The_Evolution_in_Product_ID.pdf.

EPCglobal Tag Data Standards Version 1.3. Version of March 8, 2006. www.epcglobalinc.org/standards.

Federal Communications Commission. Undated. Index of Media Bureau CDBS public database files. www.fcc.gov/mb/databases/cdbs.

Garfield, Eugene. 1961. An algorithm for translating chemical names to molecular formulas. Ph.D. diss., University of Pennsylvania. www.garfield.library.upenn.edu/essays/v7p441y1984.pdf.

Jeffrey, Charles. 1973. *Biological Nomenclature*. New York: Crane, Russak.

Jockey Club. 2003. *The American Stud Book: Principal Rules and Requirements*. Lexington, Ky.: Jockey Club. www.jockeyclub.com/pdfs/RULES_2003_PRINT.pdf.

Knuth, Donald E. 1973. Hashing. In *The Art of Computer Programming*. Vol. 3, *Sorting and Searching*. Reading, Mass.: Addison-Wesley.

McNamee, Joe. 2003. Why do we care about names and numbers? www.circleid.com/article/336_0_1_0.

Mockpetris, P. 1987. Domain names: Implementation and specification. Network Working Group Request for Comments 1035. www.ietf.org/rfc/rfc1035.txt.

NeuStar, Inc. 2003. North American Numbering Plan Administration Annual Report, Jan. 1–Dec. 31, 2003. www.nanpa.com/reports/2003_NANPA_Annual_Report.pdf.

Savory, Theodore. 1962. *Naming the Living World: An Introduction to the Principles of Biological Nomenclature*. London: English Universities Press.

Uniform Code Council, Inc. Undated. 2005 Sunrise: Executive summary. www.uc-council.org/ean_ucc_system/stnds_and_tech/2005_sunrise.html.

Chapter 10: Third Base

Bharati Krsna Tirtha, Swami. 1965. *Vedic Mathematics, or Sixteen Simple Mathematical Formulae from the Vedas (for One-Line Answers to All Mathematical Problems)*. Varanasi, India: Hindu Vishvavidyalaya Sanskrit Publication Board.

Cauchy, Augustin. 1840. Sur les moyens d'éviter les erreurs dans les calculs numériques. *Comptes rendus hebdomadaires des séances de l'Académie des sciences* 11:789–98.

Colson, John. 1726. A short account of negativo-affirmative arithmetick. *Philosophical Transactions of the Royal Society of London* 34:161–73.

Ekhad, Shalosh B., and Doron Zeilberger. 1998. There are more than $2^{n/17}$ n-letter ternary square-free words. *Journal of Integer Sequences* 1, article 98.1.9.

Engineering Research Associates, Inc. 1950. *High-Speed Computing Devices*. New York: McGraw-Hill.

Erdős, Paul, and Ronald L. Graham. 1980. *Old and New Problems and Results in Combinatorial Number Theory*. Geneva: L'Enseignement mathématique, Université de Genève.

European Museum on Computer Science and Technology. 1998. Nikolai Brusentsov—the creator of the trinary computer. www.icfcst.kiev.ua/museum/Brusentsov.html.

Frieder, G., A. Fong, and C. Y. Chow. 1973. A balanced-ternary computer. *Conference Record of the 1973 International Symposium on Multiple-Valued Logic*, pp. 68–88.

Gardner, Martin. 1964. The "tyranny of 10" overthrown with the ternary number system. *Scientific American* 210(5):118–24.

Glusker, Mark, David M. Hogan, and Pamela Vass. 2005. The ternary calculating machine of Thomas Fowler. *IEEE Annals of the History of Computing* 27(3):4–22.

Grimm, Uwe. 2001. Improved bounds on the number of ternary square-free words. arxiv.org/abs/math.CO/0105245.

Grosch, Herbert R. J. 1991. *Computer: Bit Slices from a Life*. Novato, Calif.: Third Millennium Books.

Knuth, Donald E. 1981. *The Art of Computer Programming*. Vol. 2, *Seminumerical Algorithms*. 2nd ed. Reading, Mass: Addison-Wesley, pp. 190–93.

Lalanne, Léon. 1840. Note sur quelques propositions d'arithmologie élémentaire. *Comptes rendus hebdomadaires des séances de l'Académie des sciences* 11:903–5.

Leslie, John. 1820. *The Philosophy of Arithmetic, Exhibiting a Progressive View of the Theory and Practice of Calculation, with Tables for the Multiplication of Numbers as Far as One Thousand*. Edinburgh: William and Charles Tait.

Ornstein, Leonard. 1969. Hierarchic heuristics: Their relevance to economic pattern recognition and high-speed data-processing. Unpublished manuscript available at citeseer.ist.psu.edu/ornstein69hierarchic.html.

Rine, David C., ed. 1984. *Computer Science and Multiple-Valued Logic: Theory and Applications*. 2nd ed. Amsterdam: North-Holland.

Shannon, C. E. 1950. A symmetrical notation for numbers. *American Mathematical Monthly* 57:90–93.

Sun, Xinyu. 2003. New lower bound on the number of ternary square-free words. *Journal of Integer Sequences* 6(3), article 03.3.2.

Thue, Axel. 1912. Über die gegenseitige lage gleicher teile gewisser Zeichenreihen. In *Selected Mathematical Papers of Axel Thue*, pp. 413–77. Oslo: Universitetsforlaget.

Vardi, Ilan. 1991. The digits of 2^n in base three. In *Computational Recreations in Mathematica*, pp. 20–25. Reading, Mass.: Addison-Wesley.

Chapter 11: Identity Crisis

Baker, Henry G. 1993. Equal rights for functional objects; or, The more things change, the more they are the same. *ACM OOPS Messenger* 4(4):2–27.

Bird, Richard. 1976. *Programs and Machines: An Introduction to the Theory of Computation*. New York: John Wiley and Sons.

Davis, Philip J. 1981. Are there coincidences in mathematics? *American Mathematical Monthly* 88:311–20.

Kent, William. 1991. A rigorous model of object reference, identity, and existence. *Journal of Object-Oriented Programming* 4(3):28–36.

Khoshafian, Setrag N., and George P. Copeland. 1986. Object identity. In *OOPSLA '86 Proceedings* (Conference on Object-Oriented Programming Systems, Languages, and Applications), *Sigplan Notices* 21(11):406–16.

Minsky, Marvin. 1988. *The Society of Mind*. New York: Touchstone/ Simon & Schuster.

Ohori, Atsushi. 1990. Representing object identity in a pure functional language. In *Proceedings of the Third International Conference on Database Theory*, Paris, Dec. 1990. New York: Springer Verlag.

Pacini, Giuliano, and Maria Simi. 1978. Testing equality in Lisp-like environments. *BIT* 18:334–41.

Peano, Giuseppe. 1889. The principles of arithmetic, presented by a new method. In *Selected Works of Giuseppe Peano*. Translated and edited by Hubert C. Kennedy. Toronto: University of Toronto Press, 1973.

Russell, Bertrand. 1956. On denoting. In *Logic and Knowledge: Essays, 1901–1950*, pp. 39–56. London: Macmillan.

Chapter 12: Group Theory in the Bedroom

Bernie and Phyl's Furniture. Phyl's Furniture Facts. www.bernphyl.com/ bnp/bedding_facts.asp.

Cobb, Linda. Flipping the mattress. www.diynetwork.com/diy/lv_household_tips/article/0,2041,DIY_14119_2275112,00.html.

eHow. How to care for a mattress. eHow: Clear instructions on how to do (just about) everything. www.ehow.com/how_6302_care-mattress.html.

Gallian, Joseph A. 2005. Groups in the household. *MAA Focus* 25(5):10–11.

Gardner, Martin. 1980. Mathematical games: The capture of the monster: A mathematical group with a ridiculous number of elements. *Scientific American* 242(6):16–22.

Humphreys, J. F., and M. Y. Prest. 2004. *Numbers, Groups, and Codes.* 2nd ed. Cambridge, U.K.: Cambridge University Press.

Pemmaraju, Sriram, and Steven S. Skiena. 2003. *Computational Discrete Mathematics: Combinatorics and Graph Theory with Mathematica.* Cambridge, U.K.: Cambridge University Press.

Stevenson, Seth. 2000. Going to the mattresses: How to cut through the marketing gimmicks of Sealy, Serta, and the rest. *Slate.* slate.com/id/93956.

Index

Page numbers in italic type refer to illustrations or their captions.